D1010733

The Great Global Warming Blunder

THE GREAT GLOBAL WARMING BLUNDER

How Mother Nature Fooled the World's Top Climate Scientists

ROY W. SPENCER

ENCOUNTER BOOKS

New York · London

First American edition published in 2010 by Encounter Books, an activity of Encounter for Culture and Education, Inc., a nonprofit, tax exempt corporation.
Encounter Books website address: www.encounterbooks.com

Manufactured in the United States and printed on acid-free paper. The paper used in this publication meets the minimum requirements of ANSI/NISO Z39.48-1992 (R 1997) (*Permanence of Paper*).

FIRST AMERICAN EDITION

LIBRARY OF CONGRESS CATALOGING-IN-PUBLICATION DATA

Spencer, Roy W.
The great global warming blunder : how mother nature fooled the world's top climate scientists / by Roy W. Spencer.
p. cm.
Includes bibliographical references and index.
ISBN-13: 978-1-59403-373-5 (hardcover : alk. paper)
ISBN-10: 1-59403-373-0 (hardcover : alk. paper)
1. Global warming. I. Title.
QC981.8.G56S647 2010
551.6–dc22
2009033837

10 9 8 7 6 5 4 3 2 1

Contents

Introduction & Background

CARBON FOOTPRINTS, carbon offsets, carbon taxes, carbon credits, carbon dioxide laws and regulations, cap-and-trade, going green, green energy – these terms are now part of our modern lexicon. We are told that Earth's average temperature is higher today than it has been for hundreds or even thousands of years; that humanity, not nature, now controls the climate system; that the evidence of a manmade climate crisis is everywhere; that we must drastically reduce our greenhouse gas emissions in order to save the planet. This is the new orthodoxy.

And if you have the audacity to question the world's leading climate experts on this matter? Well, you're not alone. In this book I will expose what I consider to be the Achilles' heel of the manmade global warming theory. It takes only one good piece of evidence to destroy a scientific theory, and I believe that I have come about as close to doing so as you can get in this business. I will present new evidence for two major scientific findings related to global warming and climate change. These findings could completely change the debate.

The first finding is that the climate system is much less sensitive to our greenhouse gas emissions than the experts claim it to be. This means that Earth's climate does not really care whether you travel by jet or SUV or bicycle. It also means that future global temperatures are about as likely to fall as they are to rise. This is something that most meteorologists, like me, already believe, but it has been difficult to prove because no one knew how to prove it – until now.

The second finding is that the climate system itself is probably responsible for most of the warming we have seen in the last 100 years or so. Contrary to popular belief, you don't need a change in the sun or a volcanic eruption or pollution by humankind to cause global warming or cooling. Climate change is simply what the climate system does. We now have satellite-measured evidence of this self-induced climate change: a natural mode of climate variability called the Pacific Decadal Oscillation, or PDO. Having actual measurements of the source of climate change is doing better than the theory of manmade global warming. That theory depends on forcing that is too weak to be observed even from our best Earth-monitoring satellites. It has to be computed on a theoretical basis instead.

While the evidence I will present here shows that nature causes its own global warming, I find that many people think of "global warming" as synonymous with "manmade global warming." The alleged connection between global warming and human activity has become so firmly entrenched in our minds that even after I explain the evidence that warming might be more natural than manmade, I still get questions like, "What about the melting glaciers and sea ice? Isn't that evidence of global warming?"

Arghhh . . . Warming, yes. Manmade, *no*.

In fact, the question I am asked most frequently by the public is: "Couldn't global warming just be part of a natural cycle?" And my answer to that question is yes!

Invasion of the Body Modelers

With so many other climate experts out there telling you that we are destroying the planet with our greenhouse gas emissions, why should you believe me when I disagree? To answer that question, I will illustrate my role in climate research with an analogy between climatology and human physiology.

The average temperature of the human body is 98.6 deg. F. And where does the energy come from to keep our bodies that warm? From the food we eat, of course.

Now let's suppose that everyone in the world has always consumed the same number of calories each day: 2,000. I know it sounds a little farfetched, but let's say this is a law instituted long ago by the King of the World and enforced with daily rationing of food. Then one day the King declares that he will repeal the 2,000 Calorie Law in three years. People will be free to eat as much food as they want.

Many physiologists, doctors, and medical researchers become worried that eating more food might cause our body temperature to rise, which would be dangerous to our health. They assume, with a certain logic, that if 2,000 calories a day produces a body temperature of 98.6 deg. F, then surely 3,000 calories a day will cause a higher body temperature. But this is uncharted territory. No one knows for sure what will happen, because in this story no one has ever eaten more (or less) than 2,000 calories a day.

The King asks the United Nations to convene a panel of the world's top medical experts to study the problem. The experts decide that different research groups around the world will construct computer models of how the human body functions. When these models are completed, the body modelers will run modeling experiments to see how caloric intake affects the body's temperature. The body modeling project becomes massive, with many countries participating and their governments funding the effort with hundreds of millions of dollars each year. Because the human anatomy is so complex, the project requires expensive supercomputers and hundreds of medical specialists.

Each of these researchers is an expert in how some part of the human body functions. They develop mathematical equations that eventually do a pretty good job of mimicking various subsystems in the body. Equations for the heart, circulatory system, lungs, muscles, brain, nervous system and so forth are all assembled into computer models of how the entire human body works. After three years and billions of dollars of investment, over a dozen modeling groups around the world reach the point where their computer models do reasonably well at describing the operation of an average body. They have adjusted their models to produce an average body temperature of 98.6 deg. F. While the modeling groups attack the problem in different ways, they agree that all their models put together must surely encompass all the potential outcomes for the purpose of predicting future body temperature.

The modelers then conduct experiments, gradually inputting more calories into their models to see what happens. Periodically they get together to compare their results and refine the models. Their conclusion is always the same: If people increase their food intake, their average body temperature will rise. A few of the models suggest that the temperature increase will be moderate, but others predict that it will be large enough to be dangerous and possibly deadly.

If I am a medical expert, what role do I play in this story? Well, I'm not part of the body modeling effort. Instead, I employ the latest medical monitoring devices in the laboratory to measure how the body's temperature responds under different conditions. Rather than calculate theoretically what might happen, I investigate what actually happens with real humans. Specifically, I measure how the human body reacts when it is exposed to excessive heat or is fighting off an infection, pushing its temperature above 98.6 degrees. Since the human body experiences temperature changes for a variety of reasons on a routine basis, I consider it essential that we study and understand the body's natural response to these changes.

In the course of these experiments, I discover that the body has a thermostatic control mechanism that keeps its temperature right around 98.6 deg. F. I am not the first to discover this mechanism; a few other researchers with older equipment found similar evidence years before me.

The body modelers, however, do not believe my empirical results based on actual medical observations. They assert that their models do a good job of reproducing the body's average temperature of 98.6 degrees, and their models tell them that if we start eating more food, our body temperature will rise. They vigorously defend their models against any criticism. Virtually all medical research dollars now go into body modeling. Careers and research infrastructure have been established in the field, and there are big incentives to keep the extremely complex and expensive modeling business going.

I then publish a research paper in the peer-reviewed scientific literature describing some of my early results suggesting that the body's temperature is thermostatically controlled. I also publish a paper describing how the body modelers might be fooled if they are not careful about how they interpret some very fundamental processes in the body that regulate temperature. But these articles are met by silence from the scientific community. Despite the importance of my new research to the body modeling effort and to the future of mankind's eating habits, even the news media refuse to report on the results. While my work suggests that people can eat more without having to worry about developing a fever as a consequence, the media are not interested in reporting good news. They would rather sensationalize any bad news.

Besides, everyone knows that if people can eat more than 2,000 calories a day without getting a fever, the rich will be able to eat more than the poor because they can afford more food. Many scientists, and even many citizens, feel that this will only exacerbate the inequities that already exist in society. The TV talk shows are flooded with celebrities discussing how unfair this will be to the poor.

Ultimately I find enough evidence to virtually prove my theory, but now the research papers that I submit for publication are rejected outright. In fact, one reason given for the rejection is that I am trying to publish findings that contradict the body modelers. I am getting too close to proving that they have made some fundamental errors that will invalidate their predictions for the future of the human race.

Takin' It to the Streets

The preceding story illustrates where I stand as a climate researcher today, late in 2009. The climate modelers and their supporters in government are largely in control of the research funding, which means that most government contracts and grants go toward those investigators who support the party line on global warming. Sympathizers preside as editors overseeing what can and cannot be published in research journals. Now they even rule over several of our professional societies, organizations that should be promoting scientific curiosity no matter where it leads.

In light of these developments, I have decided to take my message to the people. This message is that mankind's influence on climate is small and will continue to be small. While Al Gore likes to say that "the Earth has a fever," I will argue that the fever is natural and that it will eventually subside on its own.

This is a very different message from the one repeated *ad nauseam* in the news media: that the overwhelming consensus of scientists is that our greenhouse gas emissions have caused the global average warming of about 0.7 deg. C (1.3 deg. F) measured over the last 100 years. We are told to believe this claim because the largest body of scientists ever assembled to address any scientific issue has said it is so. The United Nations' Intergovernmental Panel on Climate Change (IPCC), formed in 1988, has relied on the participation of hundreds of governmental representatives, bureaucrats, and scientists to provide the governments of the world with scientific guidance on the issue of anthropogenic (human-caused) climate change. Their fourth and most recent report, issued in 2007, states that global warming is now confidently known to be mostly manmade.[1] "Most of the observed increase in global-average temperatures since the mid-20th Century," says the report, "is very likely due to the observed increase in anthropogenic greenhouse gas concentrations." And the IPCC must be right because they have computer models that run on big, expensive computers.

Just how much warming does the IPCC predict? Based on the twenty-something climate models involved in the effort, the 2007 report's official party line is that the total amount of warming expected to result from a doubling of atmospheric carbon dioxide is "likely to be in the range 2 to 4.5°C (3.8 to 8.1 deg. F) with a best estimate of about 3°C (5.4 deg. F), and is very unlikely to be less than 1.5°C (2.7 deg. F). Values substantially higher than 4.5°C (8.1 deg. F) cannot be excluded."

A date for the doubling of carbon dioxide is not mentioned because there are so many uncertainties about how much of it will be produced by humanity in the next 50 to 100 years. Assuming business as usual, with continued economic growth and fossil fuels dominating the global energy mix, a rough estimate is before the year 2100.

It is interesting that the predicted range of warming is not very different from what it was twenty years ago, when climate modeling was in its infancy. If we have made so much progress in computer modeling and understanding of the climate system, why is there still so much uncertainty? I believe the uncertainty stems from a fundamental misinterpretation that climate researchers have made when observing natural climate variability. This misinterpretation has found its way into the computer models that are now forecasting levels of future warming that range from significant to catastrophic. In fact, we have a peer-reviewed scientific publication that addresses the issue.[2] Unfortunately, the mainstream media have refused to report on our work. And as far as I can tell, the published evidence has largely been ignored by the scientists who should be taking notice.

More than one scientist associated with the IPCC effort has asked me, "What else could be causing the warming, other than rising carbon dioxide concentrations?" Moreover, the argument goes, if the climate system is as sensitive as many researchers believe it to be, then increasing atmospheric carbon dioxide is sufficient to explain global warming. No other reason is needed, so why should anyone bother to look for a reason other than humanity's greenhouse gas emissions?

I will be presenting evidence that the climate system is not nearly sensitive enough for the extra carbon dioxide to be the culprit. Furthermore, our latest satellite measurements of natural climate variability, combined with a simple climate model, indicate that there is an alternative, *natural* explanation for most of our recent warming. As a result of this new evidence, I will argue that a natural cause for climate change mostly eliminates the need for a human cause. After all, if the IPCC can claim that humanity's greenhouse gas emissions are all that is needed to explain global warming, then why can't I show evidence that a natural source is all that is needed to explain warming?

I hope to convince you that the IPCC has systematically ignored the 800-pound gorilla in the room: natural, internally generated climate variability, or "climate chaos." And the source of this climate chaos? Clouds.

Politicization of Science by the IPCC

The IPCC process for reviewing the science of global warming and climate change has been a peculiar perversion of the usual practice of scientific investigation. Science normally involves the testing of alternative hypotheses, not picking the first one that comes along and then religiously sticking to it. But that is exactly what the IPCC has done.

As I wrote this book, I found myself increasingly criticizing the IPCC's leadership and the way it has politicized my scientific discipline, atmospheric science, in order to promote specific policies. The truth is that the IPCC doesn't actually do scientific research. It is primarily a political advocacy group that cloaks itself in the aura of scientific respectability while it cherry-picks the science that best supports its desired policy outcomes, and marginalizes or ignores science that might contradict the party line. It claims to be policy-neutral, yet it will not entertain any science that might indicate there is no need for policy change on greenhouse gas emissions.

Contrary to what the public has been led to believe, the IPCC's relatively brief *Summary for Policymakers* – the only part of their voluminous report that a policymaker will ever read – is not written by hundreds of scientists, but by about fifty handpicked true believers who spin the science of climate change to support specific policy goals. And those goals have not changed in the twenty years of the IPCC's existence.

In the early 1990s, shortly after the IPCC was organized, President Clinton's chief environmental scientist, Dr. Robert Watson, told me that after he had helped get the production of Freon banned by the international community with the Montreal Protocol, next on the list to be regulated was carbon dioxide. There was no mention of investigating the science behind the claim that global warming was manmade–only a specific policy outcome that the IPCC was going to support. Dr. Watson later became one of the IPCC's directors, from 1997 to 2002.

The IPCC effort led to negotiation of the Kyoto Protocol to limit the production of greenhouse gases, at Kyoto, Japan, in December 1997. Those countries that later signed and ratified the Kyoto treaty are now obligated to specific reductions in greenhouse gas emissions from 2008 to 2012, after which Kyoto runs out. A new agreement for post-2012 reductions in greenhouse gases was planned for a December 2009 meeting in Copenhagen, but a global economic recession combined with protests from undeveloped and developing countries have delayed any agreement until 2010 or later.

I want to make it clear that when I criticize the IPCC, I am mostly criticizing their leadership. Those leaders are the ones who have misused science for their own political, professional, or financial gain, and then told the rest of us not to question their conclusions. Aside from their almost total neglect of the role of nature in climate change, the scientists supporting the IPCC effort have done a pretty good job of summarizing the science of global warming, along with many of the uncertainties. It is the IPCC leadership that has decided to minimize those uncertainties, and to maximize the alarm and political advocacy.

This doesn't mean there are not any concerned scientists involved in the IPCC effort; there are. But those scientists are not driving the process. As far as I can tell, the IPCC's influence and message are controlled by several dozen bureaucrats and politically active scientists who have a shared purpose and goal. The rest of the climate research community involved in the effort are just along for the ride, assured of continued funding from their respective governments on a subject of great importance to humanity. Not a bad gig for a scientist.

The primary goal of climate research is no longer the advancement of knowledge; it is instead the protection and dissemination of the IPCC party line. The peer review process for getting research proposals funded and scientific papers published is no longer objective, but is instead short-circuited by zealots adhering to their faith that humans now control the fate of Earth's climate. Scientific papers that claim all kinds of supposedly dire consequences of anthropogenic climate change are uncritically accepted and rushed to publication, while any papers that cast doubt on the premise of a human-controlled climate system are rejected.

The global warming issue has accumulated so much political and financial baggage that it will now be extremely difficult to budge the "scientific consensus" away from what a handful of bureaucrats and politically savvy scientists have decided the scientific consensus should be. As I described in my first book, *Climate Confusion*, scientists are just as prone to bias as anyone else, and when it comes to global warming it seems that everyone has biases and vested interests.[3]

The IPCC's claim that climate change is caused by human activity has led to widespread fears that if we do not act to reduce greenhouse gas emissions soon, we are all doomed. Al Gore has been the leading political proponent of this view, having received both an Academy Award for his global warming documentary, *An Inconvenient Truth*, and a Nobel Peace Prize for taking on an issue that some have apparently decided is central to world peace. The IPCC shared in that Nobel Prize as well. Mr. Gore even received a Grammy for best spoken word album, the audio version of *An Inconvenient Truth*. While Gore has falsely impugned the financial motives of scientists like me, he has made millions of dollars by actively selling the "cure" for the "disease" he claims we all have caused.[4] The hypocrisy of those who turn a blind eye to this financial conflict of interest continues to astound me.

Al Gore's leading scientific advisor on the issue, James Hansen, has also been increasingly vocal in his claim that global warming is a serious threat. Dr. Hansen, the director of NASA's Goddard Institute for Space Studies in New York City, has been modeling the climate system with computers longer than just about anyone else. He appears to be more convinced than ever that we are rapidly approaching climate "tipping points." For instance, Hansen claims that a meltdown of the Greenland ice sheet will be unavoidable if we do not start reducing our carbon dioxide emissions very soon.[5]

My Motives

Why am I willing to stick my neck out on an issue where there is so much momentum running in the opposite direction? Because the United States is making decisions on energy policy that will literally lead to death and suffering. The environmental lobby, activist news media, opportunistic politicians–and even a few Big Oil interests–have led the public to believe that we can "go green" in generating energy. But the truth is that there are still no large-scale replacements for fossil fuels that are going to make much of a difference to global carbon dioxide emissions in the foreseeable future.[6]

Should we be working on alternatives? Of course; and both government and the private sector are doing so. But all of the proposed alternatives so far are too meager and too expensive. And one of the most basic truths of economics is that when we divert resources away from more productive uses to less productive ones, people will suffer. It is usually the poor who are hurt first, and hurt the worst.

Now appearing on the horizon are energy policy changes that I fear will cause a humanitarian crisis among the world's poor. The governmental regulation of carbon dioxide emissions is expected by many to begin soon, if it has not already started by the time you read this book. The U.S. House of Representatives has passed legislation that would cap the total carbon dioxide emissions from industry and business. This would require a new bureaucracy to oversee the management, accounting, and trading of carbon emissions credits among companies. Even if this legislation stalls in the Senate, the Supreme Court ruled in 2007 that carbon dioxide is a "pollutant" and told the Environmental Protection Agency (EPA) that it must decide whether to regulate CO_2 emissions under the Clean Air Act.[7] The early indications are that President Barack Obama would support either legislation or regulation.

Too many people still do not realize that the unintended consequences of these decisions would be enormous. We have already seen corn prices skyrocket as we divert corn crops from food to ethanol production, a misguided policy that has directly hurt the world's poor. Gasoline prices have soared because we have not drilled for oil in enough places and our refinery infrastructure is too fragile. Expensive advertising campaigns by environmental groups have misled policymakers into thinking that the public opposes more drilling and refining. Even some energy companies are jumping on the bandwagon as they pander to public sentiment, misleading us by making it look like they are making great strides in green energy. Electric power companies are now being prevented from building new coal-fired plants. If they are required to use intermittent energy sources such as wind and solar power, we will eventually see brownouts and blackouts.

While relatively wealthy and environmentally conscious Westerners can deal with the higher food prices that result from diverting some of our food supply into liquid fuels, green energy policies will push many of the world's poor who are already malnourished into starvation. Many Westerners are able to absorb the extra costs of CO_2 regulation that must inevitably be passed on to the consumer, but the war on global warming will increasingly become a war on the poor.

As the United States careens toward governmental controls on energy use, citizens of the United Kingdom and the European Union have already been down this road.[8] The British were initially very supportive of restrictions on CO_2 production. But with prices for energy and other goods soaring, and little or no progress made toward the goal of reducing greenhouse gases, they are now revolting against the political establishment. Global warming is now viewed as one more excuse for the government to get its hands on the people's money.

Meanwhile, Russia's growing control over Europe's natural gas supply is a security disaster just waiting to happen.[9] As green concerns have pushed some EU countries toward more reliance on natural gas, their political future is increasingly in the hands of Gazprom and Vladimir Putin, who has been trying to buy up natural gas companies around the world–including in the United States.

Once CO_2 regulations are implemented, the price of virtually *everything* will increase, because all goods and services require some input of energy. These cost increases won't be absorbed by the energy companies, but by the consumers. If energy companies are required by law to absorb the increased costs, they will simply go out of business. The choice will come down to expensive electricity or no electricity.

If it were not for the supposed threat of global warming, Al Gore and the Supreme Court would not be able to get away with their claim that carbon dioxide is a pollutant. As most of us learned in school, atmospheric carbon dioxide is just as necessary for life on Earth as oxygen. Without CO_2 there would be no photosynthesis, and therefore no plants, and no animals, and no people either. Yet Mr. Gore has referred to our emissions of CO_2 as equivalent to treating the atmosphere like an "open sewer."[10] He and James Hansen have even called for civil disobedience to prevent the future construction of coal-fired electric power plants, which are a major source of CO_2 emissions.[11]

Another reason why I am taking my case to the people is because of its simplicity. The fundamental mistake that the climate experts have made on the science of global warming is not overly complex or obscure: they have simply mixed up cause and effect when observing cloud and temperature behavior. You could say that they have been fooled by Mother Nature. In fact, I have found that the issue of causation is one that the public understands better than the scientists do.

Comments I have received from the public over the years indicate that many of our citizens – probably a majority of them – are distrustful of the claim that global warming is manmade. In October 2008, a survey commissioned by the Nature Conservancy revealed that only 18 percent of respondents strongly believed that global warming was real, manmade, and harmful.[12] I now have evidence that the public has been right and the world's top scientists have been wrong. The importance of the global warming issue to humanity demands that the public become better informed on the reasons why so many scientists think global warming is manmade, and why they are wrong.

Unfortunately, the IPCC would have you believe that they are the only ones qualified to cast judgment on the causes of global warming.

Scientific Elitism

I claim that the theory that our greenhouse gas emissions cause global warming can be refuted with some fairly basic concepts combined with satellite observations of Earth. The evidence and arguments should be understandable to most eighth-graders.

Climate modelers will try to convince you that the only way to understand and predict global warming is with their highly complex computerized models. This allows them to claim that the evidence for manmade global warming is beyond your capability to grasp. But their work is virtually impossible to replicate because the models are so complex and the modeling effort involves lots of people at great expense. Yet being able to replicate results is a basic requirement for scientific research.

The scientific elitists who claim to speak for the climate research community have considerable disdain for the views of meteorologists, like me. I have found that most meteorologists by training are suspicious of climate models, and the modelers don't like it. For instance, there was an American Meteorological Society conference in 2008 where TV and radio broadcast meteorologists were scolded by a panel of IPCC experts who told them not to express doubts about manmade global warming on-air.[13] TV meteorologists are, after all, only meteorologists, while climate modelers are the Keepers of All Climate Knowledge.

Their complex models now supposedly constitute our main source of climate truth. Very little climate research is done anymore where scientists dig into actual observations of the climate system in order to figure out how nature works. Instead, computerized crystal balls are built and analyzed by wizards who alone are able to interpret their message for us. And just as in *The Wizard of Oz*, we are supposed to pay no attention to that man behind the curtain who is turning the knobs and pulling the levers.

But the climate modelers seem to have forgotten something that even the public recognizes: the output of computers is no better than the information that the programmers put in. As the old saying goes: garbage in, garbage out. This is not to say that climate models are garbage. I'm quite confident that if they were adjusted to agree with the satellite measurements I will be describing, their predictions of substantial global warming would largely evaporate.

I admit that the allure of theoretical models is strong. They are clean, precise, even elegant, whereas actual observations of the climate system are often incomplete, ambiguous, and open to error. There is something magical about the numbers that come out of a computer, as if they have been imbued with some divine power to reveal nature's secrets to us. But a computer is just a tool; it will do only what it is instructed to do. A scientist might be surprised with the result that the computer spits out, but that is most likely because he didn't fully understand what he was telling the computer to do.

I believe that models are necessary for determining whether our concepts of how nature works can be supported with actual numbers and known physical laws. In this book I will be using a simple computer model to interpret what nature is telling us through our satellite measurements of the Earth. Even though this model is simple enough to run in a spreadsheet program on your home computer, it is still powerful enough to study how the climate system really works.

So it isn't climate models *per se* that are the problem, but how they are used. I suppose you could say that climate models don't kill theories of natural climate change; climate modelers do.

Missing the Forest for the Trees

Climate models are built up from many components, or subsystems, each representing different parts of the climate system. The expectation of the modelers is that the greater the complexity in the models, the more accurate their forecasts of climate change will be. But they are deceiving themselves. The truth is that the more complex the system that is modeled, the greater the chance that the model will produce unrealistic behavior.

Fortunately, there is an alternative way to study complex physical systems, called emergent structures analysis. Rather then model the system from the bottom up with many building blocks, one looks at how the system as a whole behaves. The global climate system is an excellent example of an emergent structure because the operation of the whole is not obvious from how all the components work individually. In other words, even though the climate system is made up of all the individual weather systems scattered around the Earth, the way that the entire system behaves in response to some forcing is not obvious from how the individual components of the system work.

Emergent structures analysis is the kind of research that few climate scientists do anymore. I think that the modelers have missed the forest for the trees. They have been so intent on modeling individual trees in order to determine whether the whole forest will expand or shrink, that they have not bothered to examine the times when the forest actually did grow and shrink, and try to understand the reasons.

In contrast to all the IPCC's modeled complexity masquerading as scientific evidence, I will show you actual observations of how the Earth as a whole behaves. These measurements strongly suggest that the climate modelers have made a fundamental error. We will see that researchers have reasoned themselves in a circle by first assuming that natural climate change does not exist, and then building climate models suggesting that only human pollution is needed to explain global warming. This circular reasoning has led to the construction of a huge house of cards, and it's only a matter of time before the whole edifice collapses.

Forcing & Feedback (Cause & Effect)

Conceptually, there are two main processes that govern any kind of climate change: forcing and feedback.[14] These terms might sound technical, but you are already familiar with the concepts from your everyday experience. While a few climate experts will probably cringe at the analogy, these two processes may also be called cause and effect.

On the forcing side of the climate change issue, I largely agree with the IPCC. Mankind's burning of fossil fuels is slowly adding more carbon dioxide to the atmosphere. And since CO_2 accounts for a minor portion of the natural greenhouse effect that helps keep the Earth's surface "habitably warm," it is reasonable to expect that more CO_2 should cause some level of warming.

It is the feedback part of the problem where major mistakes have been made. While forcing determines whether a temperature change will occur at all, feedbacks determine just how large that temperature change will be. Positive feedbacks make the temperature change larger, while negative feedbacks make it smaller. Positive feedbacks create what we call a sensitive climate system, while negative feedbacks correspond to an insensitive climate system.

If the climate system is very sensitive, then the small warming tendency from increasing atmospheric concentrations of CO_2 will be amplified. This is the IPCC's position on feedbacks. In a sufficiently sensitive climate system we can explain most if not all global warming to date with humanity's greenhouse gas emissions alone. Furthermore, a sensitive climate system would also mean that we can expect significant manmade global warming to continue – maybe even accelerate – into the future. Scientists' belief in a sensitive climate system explains why you keep hearing about the dangers of methane emissions from cows and other seemingly innocuous forcings. If the climate system is highly sensitive, then we have to worry about many sources of greenhouse gases and particulate pollution.

But if the climate system is relatively insensitive to forcing, then the extra carbon dioxide in the atmosphere cannot explain the warming we have observed. There must be some stronger, natural warming mechanism at work. An insensitive climate system will not really care how much methane has been produced by the time you eat your hamburger, or whether you drive a huge SUV. An insensitive climate system resists temperature change – not preventing it entirely, but reducing its magnitude.

If I am correct in regarding the climate system as insensitive, then the twenty computerized climate models being run in several countries around the world are predicting far too much global warming. If we are not causing global warming, then reducing carbon dioxide emissions to "fix" the problem will have no measurable effect on global temperatures.

The research community's confusion of forcing and feedback – cause and effect – is a major theme of this book. In particular, the role of causation in cloud behavior is at the core of what I believe to be the greatest scientific *faux pas* in history. The mistake that researchers have made can best be introduced in the form of a question: When the Earth is observed to warm, and cloud cover decreases with that warming, did the warming cause the clouds to decrease, or did the decrease in clouds cause the warming? In the big picture of climate change, cloud changes causing temperature changes would be called forcing, while temperature changes causing cloud changes would be called feedback. Both occur in nature all the time. Yet when researchers have estimated feedbacks by analyzing natural climate variations, they have assumed causation in only one direction.

Because researchers have not accounted for natural cloud fluctuations forcing temperature variations, the illusion of a climate system dominated by positive feedback has emerged. I had always suspected that researchers were mixing up cause and effect even before I got into this line of research, but until recently I was not able to prove it.

What I am claiming is more than just an untested hypothesis; my colleagues and I have published papers in the peer-reviewed scientific literature that have been laying out the evidence step by step.[15] But chances are you haven't heard about our work. This is because the mainstream media are not interested in covering any news stories about climate that do not support Al Gore's apocalyptic vision of a global warming Armageddon. Other scientists have had similar experiences with their published research. As a friend from a newspaper family once told me, "bad news is good news, and good news is no news."

When I have talked to reporters about our published research, they either ignore our results or find another scientist who will dismiss my views without knowing what I'm talking about. Or, more often, they do not even contact us in the first place. After all, how could the consensus of hundreds of the world's best scientists be wrong? And why would any reporter want to interview a scientist who is painted as the equivalent of a "Holocaust denier"[16] anyway?

One of the problems with climate research is that most researchers are so specialized that they either have no interest in reading your research publication, or do not understand the implications of what you have presented. So, even if you publish research that does not support the belief in a sensitive climate system, most other researchers will be either unaware of your work or unable to figure out how your results fit into the global warming "big picture."

* * *

Manmade Climate Change – or Natural?

If the Earth's climate is largely insensitive to our greenhouse gas emissions, then what has caused the warming we have experienced over the last 100 years? If our greenhouse gas emissions are too weak to have caused it, there must be some stronger, natural forcing at work.

I will advance the argument that *natural, internally generated cloud variability* is responsible for most of the climate change we have seen up to the present and will likely see in the future. And contrary to the claims of some scientists that recent warming is unprecedented, the warming we experienced through the twentieth century is not much different from that experienced during other centuries over the last 2,000 years.

At this point you might be thinking, "Well, of course natural climate change happens." But this has been surprisingly difficult to prove scientifically. The IPCC avoids the subject because it detracts from the claim that humans are now the main driver of climate. As we will see, the IPCC has even attempted to eliminate the Medieval Warm Period and the Little Ice Age, two events that we know from the historical record actually occurred.

The IPCC scientists proclaim confidence that their climate models are behaving realistically and can explain global warming by anthropogenic pollution alone. But does one hypothesized explanation remove the need to search for alternative explanations? What if there are other explanations that fit the observations better? After all, alternative hypotheses are fundamental to the practice of science. Competing scientific explanations sharpen our understanding and help us arrive at a more accurate explanation of how the physical world works.

Except, apparently, when the subject is global warming.

Another Global Warming Book?

Most books on global warming deal with a bunch of little pieces of a huge puzzle. I will instead address the single most important piece, the one that determines what the finished puzzle looks like: feedbacks.

I find it difficult to believe that I am the first researcher to figure out what I describe in this book. Either I am smarter than the rest of the world's climate scientists – which seems unlikely – or there are other scientists who also have evidence that global warming could be mostly natural, but have been hiding it. That is a serious charge, I know, but it is a conclusion that is difficult for me to avoid.

For those who have read my first book, *Climate Confusion*, this book contains new and important science that supports my view that the Earth is much more resilient than most scientists claim. You might say that, rather than "hot, flat, and crowded," I believe the Earth to be cool, round, and spacious. I hope this book will lead to a better-informed public that can more critically evaluate the claim that adding carbon dioxide to the atmosphere is a menace to life on Earth. Whether carbon dioxide regulations or laws are still being debated as you read this or have already been implemented, you will be better equipped to influence the political process and to help prevent or rescind misguided and dangerous laws or regulations on the production of carbon dioxide.

I also hope to spur other scientists to investigate my claims on their own, and to speak out if they agree with me that the last few decades of myopic global warming research has resulted in the greatest scientific blunder in history. I don't know whether it will take two years or twenty, but I predict that at some point in the future we will realize that the fear of catastrophic climate change was the worst case of mass hysteria the world has ever known.

Chapter 1 · Climate Change Happens

Despite what the United Nations' IPCC would like you to believe, natural climate variability occurs on every time scale of any practical interest to humans: years, decades, centuries, millennia, and everything in between. Some of this variability is due to known cycles such as El Niño, La Niña, and the Pacific Decadal Oscillation. What the "scientific consensus" has failed to account for is that global warming (or cooling) can happen through natural cloud changes altering the amount of sunlight being absorbed by the Earth.

You wouldn't think that a book on climate change would need to prove that natural climate variability exists. But one of the fundamental tenets of the current "scientific consensus" on global warming is that humans now control the future course of the global climate system.

The United Nations' Intergovernmental Panel on Climate Change does acknowledge that there is natural climate variability on a year-to-year basis, and maybe even decade-to-decade. After all, we have clear evidence that events like El Niño and La Niña cause some years to be warmer than others. Yet the IPCC refuses to accept that global warming (or cooling) on time scales of thirty years or more can also be caused by Mother Nature. That, apparently, is humanity's job.

But, contrary to the claims of the IPCC, there is no basis for assuming that natural climate change can't occur on just about any time scale. For instance, let's examine the last 2,000 years of global average temperature variations.

GLOBAL TEMPERATURE VARIATIONS: 0 A.D. TO 2009

The top panel of Fig. 1 shows an average of eighteen non-tree-ring temperature "proxies" from fifteen locations around the world.[1] Proxies are indirect methods used to estimate temperatures in the distant past, before there were thermometers. Tree-ring proxies were specifically excluded by the researcher who published these data since they are not very good indicators of temperature change – an issue I will return to later.

The plotted values in the first panel are thirty-year averages. The two most prominent features are the Medieval Warm Period, centered around 1000 A.D., and the Little Ice Age, which occurred several hundred years later. During the Medieval Warm Period, the Vikings arrived in Greenland and started farming. Wine grapes were being grown in England. Then, as the Little Ice Age was advancing centuries later, the Viking colonization of Greenland ended when crops failed from the long, slow slide into a colder climate.[2] In the depths of the Little Ice Age, winter carnivals ("frost fairs") were held on the frozen River Thames in London.[3] The Thames no longer freezes in winter, and the last frost fair was held during the winter of 1814.

Superimposed on these two major features are shorter periods, about 50 to 100 years in duration, when rapid temperature changes occurred, both cooling and warming. Note that the twentieth century was one of these periods of relatively rapid temperature change. This suggests that the warming in the twentieth century, while noteworthy, was not unprecedented. In fact, it appears that periods of 50 to 100 year of rapid warming or cooling have been the rule, rather than the exception, over the last two millennia.

In the second panel of Fig. 1 we zoom in on the most recent 100 years, the period during which humans are allegedly responsible for global warming. The temperature curve is now made up of five-year averages, rather than thirty-year averages, and is based on real thermometer measurements.[4] While the thermometers are sparsely distributed around the world, this at least

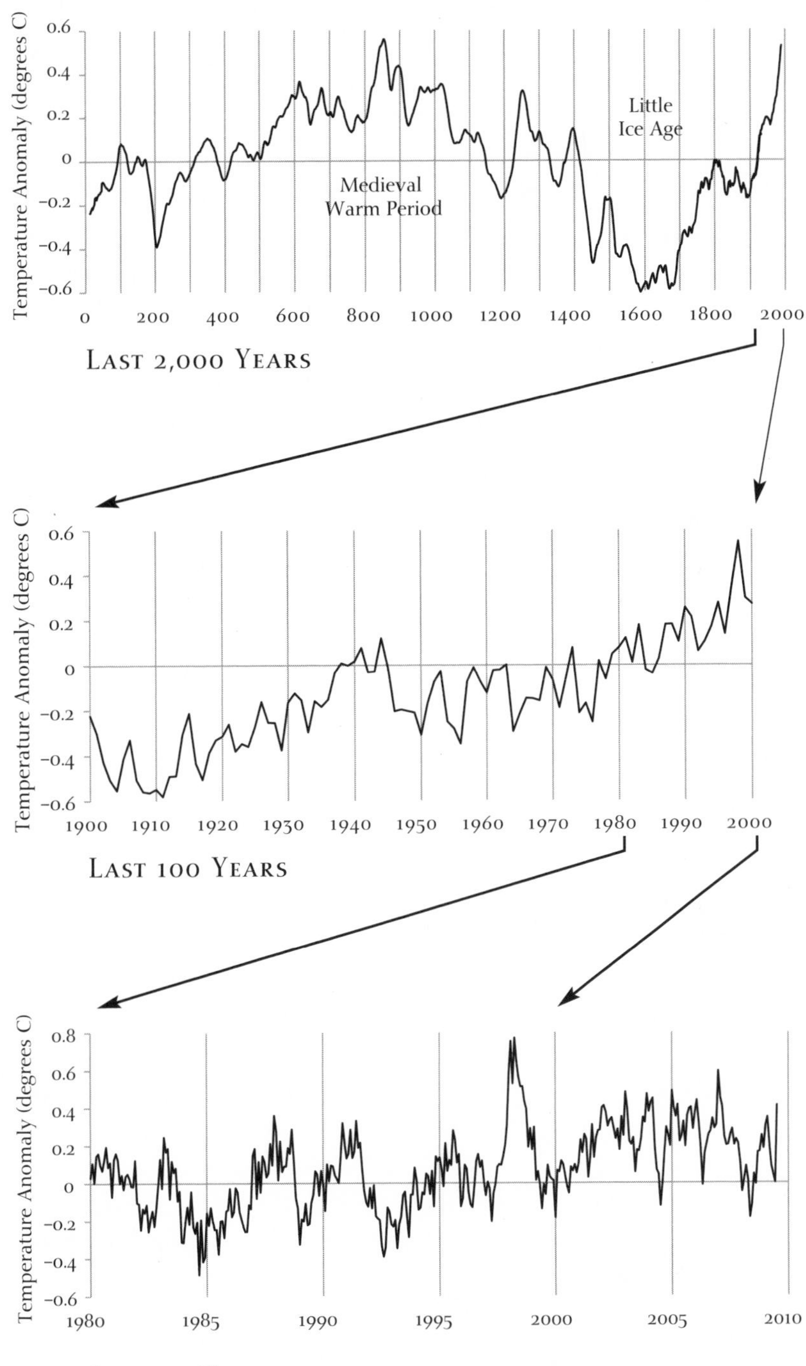

Fig. 1. Global average temperature variations over the last 2,000 years.

puts us a step closer to being able to monitor variations in global average temperature accurately.

The main feature we see over the twentieth century is an overall warming trend of about 0.7 deg. C (1.3 deg. F) per century. If we look more closely, this 100-year period appears to be split into three roughly equal segments: warming from 1900 to 1940, then slight cooling until the late 1970s, and finally resumed warming since then. Climate modelers have spent hundreds of millions of dollars over the last twenty years trying to explain this temperature behavior through human causes, mainly greenhouse gases and aerosol (particulate) pollution. In Chapter 6 we will examine the natural mechanism that I now believe is responsible for most of these temperature changes.

If we focus on the last third of the twentieth century, shown in the bottom panel of Fig. 1, we enter the satellite period of record, which allows us to make truly global measurements.[5] Instead of near-surface air temperatures, which the thermometers monitor, the satellites measure the average temperature of deep atmospheric layers. While there can be some significant differences between surface and deep-layer temperature variations over the course of several weeks, on time scales of several months or more they are tightly coupled by atmospheric convection mixing the solar heating of the Earth's surface throughout the lower atmosphere. In other words, deep-layer lower atmospheric temperature changes closely follow surface temperature changes on time scales of a few months or longer.

The geographic coverage of the Earth by the satellites is so complete that we can now calculate global average temperature variations with high precision – to about one or two hundredths of a degree C per month. We know that the large month-to-month temperature variability seen by the satellites since 1979 is real because different satellites in different Earth orbits show the same features.

All this temperature variability on a wide range of time scales reveals that just about the only thing constant in climate is change. This makes the identification of an "average" climate state very difficult, and "normal" climate nearly an oxymoron. As a result, the zero lines in the three panels of Fig. 1 are all different and somewhat arbitrary. They are based on different base periods of time chosen to reference the temperature "anomalies," or departures from average. Note that in climate monitoring we are mainly interested in changes of temperature with time, so we seldom mention the absolute temperature values. We probably do not know the average surface temperature of the Earth to better than one degree, but with satellites we can monitor temperature changes to about a hundredth of a degree. I am often asked what those averages have been in our most recent period of record. The global average near-surface air temperature has been estimated to be around 14 deg. C (57 deg. F), while the satellite-measured lower atmospheric layer averages about -4 deg. C (24 deg. F).

Most of the temperature fluctuations seen since 1979 are due to El Niño, La Niña, and two major volcanic eruptions: El Chichón in Mexico in 1982, and Mt. Pinatubo in the Philippines in 1991. During an El Niño event, the tropical Pacific Ocean becomes warmer than normal, with less upwelling of cold water off the western coasts of North and South America.[6] Normal atmospheric flows spread that unusual warmth throughout the tropical atmosphere, and then to most regions outside the tropics. The opposite happens during La Niña, with increased upwelling of cold water from the deep ocean eventually causing unusually cool global average temperatures.

On occasion, a major volcanic eruption, like Mt. Pinatubo in 1991, can eject millions of tons of sulfur into the stratosphere.[7] This sulfur is converted into sulfuric acid aerosols, which then reflect back to outer space a few percent of the sunlight that would normally have warmed the surface. One or two unusually cool summers can ensue before those volcanic aerosols gradually dissipate. It is believed that the eruption of Mt. Pinatubo caused the cool conditions of 1992–1993, as seen in the bottom panel of Fig. 1.

This volcanic cooling effect is the basis for a proposed geoengineering solution to global warming. It involves transporting massive amounts of sulfur up to the stratosphere, where it would be dumped to mimic the cooling effects of a major volcanic eruption. This "solution," of course, assumes that there is an anthropogenic global warming problem to begin with.

THE ELUSIVE "TEMPERATURE TREND"

I frequently hear the question, "Is global warming happening now?" Unfortunately, the large amount of temperature variability seen in Fig. 1 makes that question surprisingly difficult to answer. If globally averaged temperatures were steadily increasing year after year, we would be able to answer, "yes." Or, if temperatures were the same, year after year, we would be able to answer, "no."

But the huge amount of variability seen in Fig. 1, on all time scales, means that "warming" is in the eye of the beholder. One commonly heard statistic is that global cooling has been in progress since 1998. But 1998 was a particularly warm El Niño year, so that statement is quite misleading. You could also say that considerable global warming has occurred since 1999, which was a cool year. But that statement would be equally misleading.

I think the best answer is that, as of this writing in late 2009, it has not warmed since about 2001. So one might legitimately claim that "global warming stopped" in 2001. But this statement has no predictive value whatsoever, since warming could resume at any time. And because there is so much year-to-year variability, we will probably have to wait several more years before we know whether warming is "happening now." In effect, we will be able to identify warming only when we see it appear in the rearview mirror. So, there is no way to know whether global warming is happening now or not.

The warming that the IPCC considers manmade is that which occurred in the latter half of the twentieth century. In the case of the thirty-year period since 1979, for which we have satellite measurements, an underlying warming trend of about +0.13 deg. C per decade (+0.23 deg. F per decade) can be computed. Thermometer measurements from this period indicate a somewhat larger rate of warming. While this doesn't seem like a very big number, in climate terms it is regarded as fairly rapid warming. This most recent period of warming is shown as the very last up-tick in temperature plotted as a dotted line in the top panel of Fig. 1, which indicates that it rivals the strongest warming events of the last 2,000 years.

As strong as this recent spurt of warming is, it still amounts to only about one-half the IPCC's predicted rate of future warming in the twenty-first century: +0.3 deg. C per decade (0.5 deg. F per decade). This means that the IPCC expects warming to accelerate during this century, a rather bold prediction to say the least.

I hope that you now have a better understanding of why there are so many seemingly conflicting news reports, like "global warming is accelerating" or "global warming has stopped." Chances are that most of these statements contain an element of truth; they just refer to different periods of time. The confusion arises because there is so much natural variability in the climate system, on all time scales. Given all this natural variability, are we to believe that humanity is now in control of climate, as the IPCC claims?

Apples, Oranges, and Errors

Up to this point I have assumed that the global temperature estimates in Fig. 1 are free from errors. But there has been considerable debate over the accuracy of all methods of monitoring temperatures: proxies, thermometers, and satellites. No physical measurement is free of errors, and estimates of global average temperatures are no different. Some scientists have even claimed that there is no such thing as a global average temperature, and that even if there were it would be irrelevant for climate anyway. I disagree. While scientists might never agree on exactly what temperatures would go into such an average, the fact remains that the global distribution of atmospheric and surface temperatures is the largest single influence on how fast the Earth continuously loses radiant energy to outer space in the face of its continuous absorption of energy from the sun.

The temperature proxy data have been the most controversial because they are indirect, based on such things as sea sediments and stalagmites in caves, mostly in the Northern Hemisphere. There is simply no way to determine how accurate past temperature reconstructions based on proxies are. That would require many centuries of accurate thermometer measurements, and those do not exist.

Even if the proxies provided totally accurate temperature estimates, the low time resolution of the proxy estimates in Fig. 1 (thirty-year averages) must be considered before jumping to conclusions about record warm years. For instance, 1998 is generally regarded as the warmest year for global average temperatures in at least the last 150 years. A few scientists have even proclaimed 1998 to be the warmest in the last 2,000 years, if not longer.

But I consider any such statements to be meaningless, like comparing apples to oranges. The proxy data are not good enough to tell us just how warm *individual* years were, say, during the Medieval Warm Period. So, for example, there is no way to know how much warmer or cooler the year 855 A.D. was compared with the year 854 A.D.

If those individual years are embedded in a very warm thirty-year period, it is entirely possible that one or more of them was considerably warmer than the "record" year of 1998. We had daily global measurements from multiple Earth-orbiting satellites in that year, and therefore we have a very good estimate of how much warmer 1998 was than 1997, probably to a precision approaching 0.01 deg. C. But there is no way to know with confidence whether 855 A.D. was warmer than 854 A.D. It is entirely reasonable to suppose—but impossible to prove—that one or more years in the Medieval Warm Period were warmer than 1998. It is easy for scientists to make grand claims when there is no way to prove them wrong.

In fact, the time scale of the temperature proxies in Fig. 1, thirty years, is exactly the same as that used by the National Weather Service to determine climatological averages, or "normals." So, what is regarded as the highest time resolution in the proxy data is the same as the time resolution used to define climatological normal temperatures in the modern instrumental period of record. This further illustrates the absurdity of comparing the warmth of recent years with past centuries when we did not have sufficient measurements to compute accurate global averages on a yearly basis.

The Hockey Stick

As mentioned earlier, the temperature proxies in the top panel of Fig. 1 do not include any estimates from tree rings. In contrast, the famous "hockey stick" reconstructions of global temperatures over the last 600 years or more were based mostly on tree-ring measurements, particularly from Colorado.[8]

A veritable poster child for manmade global warming, the original hockey stick (labeled MBH98 in Fig. 2), and its successor (MJ03),[9] virtually eliminated evidence of the Medieval Warm Period and the Little Ice Age, making the warming since the 1800s appear even more dramatic. While the most recent incarnation of this dataset (M08 in Fig. 2)[10] shows the Medieval Warm Period and the Little Ice Age once again making an appearance, the earlier and more widely disseminated versions of the dataset had minimized the signatures of those natural climate fluctuations.

The importance of the early versions of the hockey stick to the history of the global warming debate cannot be overstated. It minimized natural climate variability over the last 1,000 to 2,000 years, thus making the warming in the twentieth century seem unprecedented.[11] And since the IPCC has maintained that the twentieth-century warming was due to mankind's greenhouse gas emissions, this made it look as if humans, not nature, control the climate system.

In my view, this was a deliberate ploy by the IPCC leadership in their effort to build the case for manmade global warming. As Chris Horner noted in his recent book, *Red Hot Lies*, one of the IPCC's lead authors once declared to a colleague in an email, "We have to get rid of the Medieval Warm Period."[12] And for the better

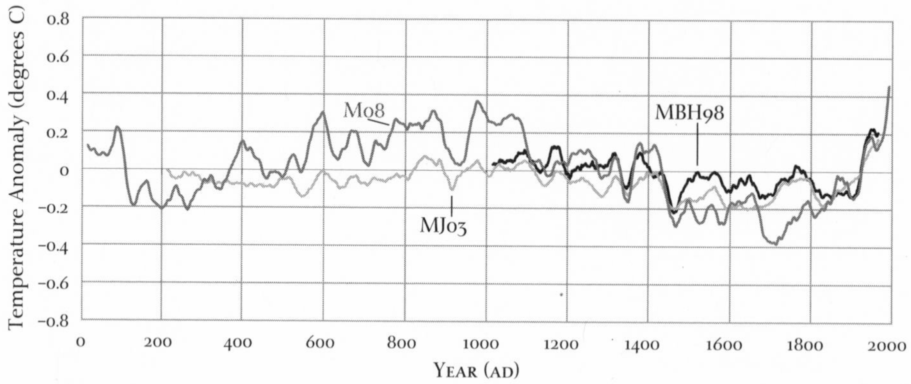

Fig. 2. Three versions of the "hockey stick" temperature reconstruction published by Michael Mann and co-authors.

part of a decade, the IPCC essentially did just that. It was stunning how swiftly and uncritically the IPCC embraced the hockey stick. Many years of published research supporting the existence of the Medieval Warm Period and the Little Ice Age were suddenly swept aside to make room for a revisionist climate history where there is no natural variability anymore, and where humans are in almost total control of the Earth's climate.

The hockey stick later fell on hard times, though. It was largely marginalized in the fourth and latest report of the IPCC, published in 2007, because an independent review by the National Academy of Science of the statistical analysis involved in the creation of the hockey stick led to the determination that it was the result of a flawed methodology.[13] This expert review might never have happened had not two Canadians, one an economist and the other a retired statistician, dug into exactly how the hockey stick was created.[14]

It's not clear that tree rings can be trusted to reconstruct past temperatures to any useable level of accuracy anyway. It turns out that the most recent tree-ring data do not even show the warming that occurred in the second half of the twentieth century, but appear to indicate a cooling instead. This discrepancy is called the "divergence problem." Craig Loehle argues that tree-ring data cannot be used as temperature proxies for previous warm events, such as the Medieval Warm Period, because the tree rings have not demonstrated sensitivity to unusual warmth in the late twentieth century.[15] In other words, tree-ring data have probably underestimated the magnitude of previous warm events in history – which is exactly what the hockey stick did – since those tree rings do not even show the warming over the last fifty years! Therefore, it could be that temperatures in the Medieval Warm Period were considerably higher than today. We simply do not know, and probably never will know.

Thermometer data, available since 1900 or earlier, are clearly better than temperature proxies, but they are still rather limited in their geographic sampling. Until recently there have been very few measurements over two-thirds of the Earth: the oceans. Early measurements of ocean temperatures were taken from buckets dipped in the ocean from the decks of ships. Later, temperatures would be taken well below a ship's water line, in the intake ports for water that cooled the ship's engine. Most recently, a global network of over a thousand drifting buoys has been deployed specifically for measuring sea surface temperature, salinity, currents, and weather. The differences between these various observing systems scattered through time mean that our estimates of ocean warming to a fraction of a degree over the last hundred years are, at best, uncertain.

Fortunately, the global sampling problem is believed to be minimized for long-term trends. Weather patterns move around the globe, generally from west to east, and this tends to average out localized warm or cold events over time, in effect "smearing" them around the Earth. Because the oceans represent such a huge reservoir of heat energy, land temperature changes tend to follow ocean temperature changes over the long run. Therefore, long-term temperature trends over land are believed to be largely a response to long-term changes in the oceans.

In fact, it has recently been demonstrated that if the oceans warm for any reason, global land areas can warm even more.[16] This makes the oceans a potential key player in long-term climate change. It also means that what the scientific consensus views as a "fingerprint" of anthropogenic warming – stronger warming over land than over ocean – is also consistent with natural climate change. The same is true of the missing tropical upper tropospheric "hot spot" that was expected to result from manmade warming.[17] It, too, would be expected to accompany a natural warming of the oceans. I believe that there is no unique fingerprint of anthropogenic global warming. Warming is warming, and it will look basically the same no matter what causes it.

One would think that land-based thermometers should provide accurate estimates of temperature changes over the years, but there are substantial sources of error. The most significant of these is the fact that thermometers are usually placed where people live, and people tend to build things, replacing native vegetation with structures, roads, parking lots, and other manmade sources of heat. While vegetation tends to cool the air by diverting some of the sun's energy into photosynthesis and evapotranspiration, most manmade surfaces and structures just sit in the sun and bake. When more people and more buildings and roads occupy the environment around a thermometer site, the air is heated in the immediate vicinity. This leads to an "urban heat island" warm bias, which typically increases with time as more buildings and parking lots are added. But the urban heat islands have virtually no direct warming effect on the rest of the Earth, since the coverage of the Earth by cities and towns is at most 1 percent.[18]

In fact, some of the thermometer sites used for climate monitoring have recently been revealed to be contaminated by heat exhaust from air conditioners and heat coming off the roofs or walls of buildings. All these influences add up to a component of the measured warming trend that is entirely local and that spuriously inflates our estimates of global warming.

While the thermometer dataset developers claim they have removed this spurious source of warming, there is increasing evidence that much of it remains in the data. One recent estimate is that as much as 50 percent of the warming measured over land in the last thirty years could be spurious, due to various indirect effects of economic growth contaminating the thermometer data.[19] If you have not heard about that important study, you have the news media to thank, since they decided it was not worthy of being reported.

Satellite instruments provide our only truly global source of temperature information. The measurements are based on either infrared or microwave emissions given off naturally by the atmosphere, and they require carefully calibrated instrumentation to provide a stable long-term record. The instruments carry their own laboratory-calibrated electronic thermometers to convert their measurements of the atmosphere into temperatures.

Probably our biggest headache in trying to monitor climate trends with satellites is not related to calibration but to the kind of Earth orbit the satellites are in. At least until recently, all the satellites had local observation times that slowly changed over the years. This would be like trying to determine climate trends from your backyard thermometer by taking measurements at noon one year, then at 1:00 P.M. the next year, then 2:00 P.M. the next, and so on. The day-night temperature cycle gets mixed in with whatever climate variability there is, and so it must be estimated and removed.

This problem has been alleviated only since mid-2002 with the launch of NASA's Aqua satellite, which carries extra fuel to adjust its orbit periodically and so maintain a constant observation time, year after year. This is the primary source of temperature data that John Christy and I use for monitoring global temperature trends.

It is clear that a host of problems are involved in the determination of temperature trends and other statistics. All our measures of temperature variability are imperfect. And even if they were perfect, the huge amount of natural variability in the climate system—on time scales from yearly to millennial or longer—makes the definition of a temperature "trend" very difficult. While we can probably say with high confidence that the climate has warmed in the last 50 to 100 years, it is more difficult to say by exactly how much, still more difficult to say whether it is unprecedented or not, and impossible to say what any of this means for future temperatures.

Next, we will examine one mode of natural climate variability that I believe plays a crucial role in what is popularly known as "global warming."

THE PACIFIC DECADAL OSCILLATION

It should be obvious by now that nature causes all kinds of climate variability, even if we have difficulty measuring it accurately. I think of this as chaos in the climate system. Chaos refers to complex, internally generated climate variations that are not well understood or predictable. These variations may appear random, but they can also exhibit some degree of regularity.

The classic example of chaos is associated with day-to-day weather. Many years ago, Ed Lorenz discovered chaotic behavior when he was experimenting with one of the first computer models being developed to predict weather.[20] He found that vanishingly small influences – presumably even the flap of a butterfly's wings – can completely change global weather patterns in a matter of months. This chaos results from what is called "sensitive dependence on initial conditions."

Chaos is the fundamental reason why weather forecasting beyond about seven or ten days has very little skill. The weather measurements we put into computerized forecast models are sparsely spread around the globe, coming mostly from weather balloon sites, surface weather stations, and commercial aircraft. Lots of atmospheric variability in between weather observation sites is never sampled, which means it never makes its way into the forecast models. This is not much of a problem for forecasts of one to three days, but the further out in time one tries to forecast weather, the larger the effect of all that unmeasured variability becomes.

We are beginning to understand some kinds of chaotic behavior in the climate system, partly because they occur with some regularity. For instance, El Niño and La Niña events come around every few years. But they are not well enough understood to predict which years will experience an event. When you do hear a forecast of El Niño or La Niña conditions for the coming months, it is only because an event has already started. The forecasters are just extrapolating what is happening today into the future since they know that these events take a year or two to run their course.

My claim is that this chaotic behavior also occurs on much longer time scales. The chaotic climate fluctuation that I highlight in this book is known as the Pacific Decadal Oscillation, or PDO.[21] Somewhat like El Niño and La Niña in the tropical Pacific, the PDO is a regional shift in weather patterns, but over the North Pacific Ocean. And rather than having a time scale of only a couple of years, the PDO changes phase much more slowly, every thirty years or so. This makes it a potential player in global warming.

El Niño, La Niña, and the PDO can be thought of as alternative ways that nature has to move heat around the Earth. The most fundamental function of both the oceanic and the atmospheric circulations is to transport heat around the globe, from regions where excess solar heating occurs to regions where there is less solar heating. Since the atmospheric circulation is coupled to the ocean circulation, a change in one is almost always accompanied by a change in the other as they work together to move heat energy from where there is more to where there is less.

The sea surface temperature anomalies (deviations from normal) for the warm and cool phases of the PDO are shown in Fig. 3.

While the importance of the PDO to the global warming debate has been largely ignored, its thirty-year time scale is long enough to cause climate change. This is comparable to the period in which the IPCC claims to have evidence of mankind's fingerprint on climate. To remind you, a major conclusion of the IPCC's 2007 report was: "Most of the observed increase in global-average temperatures since the mid-20th Century is very likely due to the observed increase in anthropogenic greenhouse gas concentrations." So, the IPCC is claiming to be confident that warming in

PACIFIC DECADAL OSCILLATION

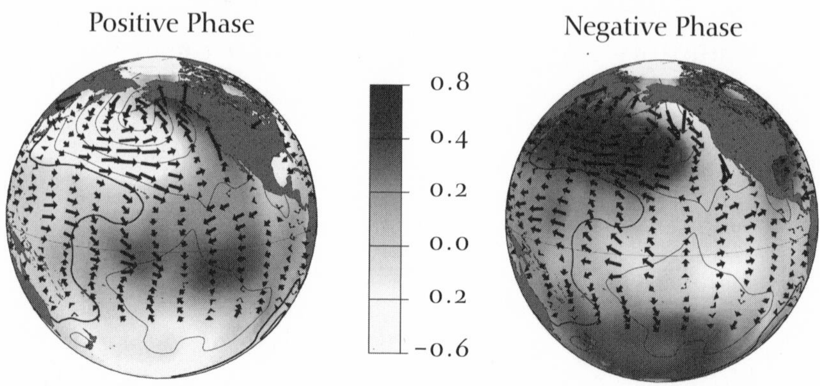

Fig. 3. The Pacific Decadal Oscillation (PDO) is a regional shift in weather patterns that changes phase every thirty years or so. The patterns are sea surface temperature (SST) anomalies (deviations from normal), and the arrows represent surface wind anomalies.

the last fifty years is manmade, and since there was slight cooling from the 1940s to the late 1970s, they are mostly referring to the warming over only a thirty-year period.

Guess what? As shown in Fig. 4, that thirty-year interval just happens to be the same period when the PDO was in its "positive" or warming phase.

Now, one would think that the IPCC reached its conclusion that mankind very likely caused the recent warming after ruling out natural climate variability, like that associated with the PDO, as a cause. But the truth is that they never seriously investigated it. The IPCC has taken for granted that there are no natural variations in global average temperatures once one gets beyond a time scale of ten years or so.

Specifically, the IPCC's most important (and incorrect) assumption is that the average cloud cover of the Earth always remains the same. It is well known that the primary role of clouds is to cool the Earth,[23] and so any long-term change in clouds is a potential source of global warming or cooling. The 2007 IPCC report does indeed mention the Pacific Decadal Oscillation and

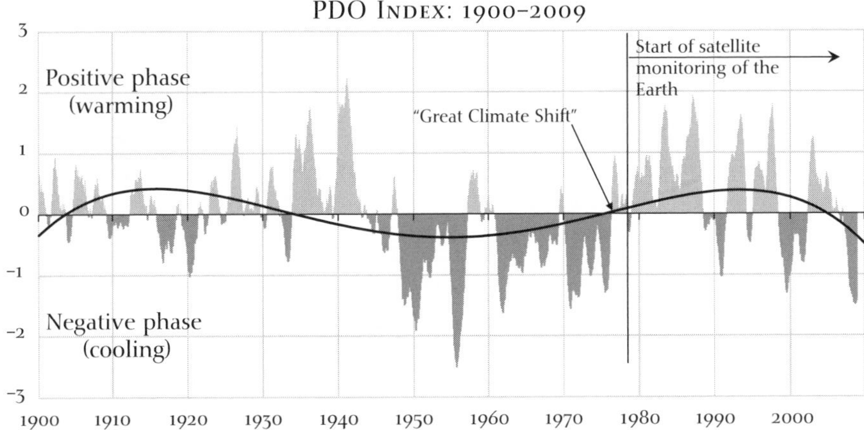

Fig. 4. Running five-month average values of the PDO index for 1900–2008.[22]

other types of multidecadal variability, but for some reason never asks the obvious question: Could these natural climate fluctuations cause a change in global cloudiness?

The assumption that global average cloudiness never changes has been rather easy for the IPCC to make since we do not have sufficient global observations with which to document, let alone understand, such chaotic fluctuations in climate. We would need highly accurate, global measurements of the climate system for at least fifty years to prove or disprove the existence of natural mechanisms of climate change. Unfortunately, those measurements do not exist. Consequently, the IPCC can correctly claim there is virtually no published research to support natural sources of long-term climate change.

This is not arguing from the evidence, though, but from a lack of evidence. Because we do not have highly accurate satellite observations of clouds, temperature, and the like over the last fifty years or more, any evidence for natural sources of climate change will require some digging to find. It's like trying to solve a murder mystery when one has very little to go on initially. Since a possible weapon (greenhouse gas emissions) was found at the scene of the crime (warming), that's good enough for the IPCC to pin the rap on humanity.

One might think that the IPCC would have thoroughly investigated natural sources of climate change. But governmental funding of climate research in recent years has been channeled primarily into gathering circumstantial evidence to connect our greenhouse gas emissions to global warming. This is why you now hear every change in nature being attributed to manmade global warming. Scientists write proposals to receive government funding to study the effect that anthropogenic global warming has had on any number of natural phenomena. Is it any surprise if they find what they were paid to find? While a few government contracts, such as my own, are worded in sufficiently general terms to allow the investigation of natural sources of climate change, most are funded mainly because they support the current global warming paradigm.

I have to wonder: what might we have found if just 10 percent of those research dollars went instead into the specific study of natural sources of climate change?

I want to emphasize again that there is no unique fingerprint of manmade global warming. Any warming due to manmade greenhouse gas emissions looks the same as warming due to, say, increasing water vapor (Earth's main greenhouse gas) resulting from a warming of the oceans. The ocean warming could, in turn, be the result of low clouds changing how much sunlight is absorbed by the oceans, or by a change in how fast cold water wells up from the ocean depths. Therefore, while patterns of warming across the Earth in the last fifty years might be "consistent" with manmade global warming, they are equally consistent with natural sources of warming.

The PDO is usually mentioned in terms of its regional impact, such as its strong influence on snow pack in western North America, or on fisheries production off the West Coast of the United States. Little formal research has been done into its possible influence on long-term climate change and global warming. Yet it is obvious from Fig. 4 that the PDO involves the longer time scales–thirty years or more–that are relevant to the global warming issue. How could a small change in weather patterns over the North Pacific Ocean cause a change in global temperatures? Well, if the split-second flap of a butterfly's wings can change global weather patterns six months later, how much more could a thirty-year change in weather patterns over the Pacific Ocean cause a change in global climate over fifty years, one hundred years, or even longer?

Again, the most likely climate impact of a change in weather patterns is its potential to change global average cloud cover. This would alter how much solar energy the oceans absorb, which would then cause global warming or cooling. This is the natural mechanism that I find the public understands and appreciates better than the climate "experts" do.

It turns out that the history of the PDO during the twentieth century shown in Fig. 4 is closely related to major climate events that you have been led to believe are caused by our pollution. The last time the PDO changed phase was in 1977, an event that some have called the "Great Climate Shift of 1977." This event brought an end to the slight global cooling trend that started in the 1940s (see Fig. 1), which was then replaced with a warming trend from the late 1970s through the 1990s.

After the Great Climate Shift, Alaska warmed immediately and then remained warm. Temperatures in the Arctic started rising, a slower process because it takes time for the ocean and sea ice to respond to a warming influence. Arctic sea ice cover was observed to start shrinking in the 1980s by our new satellite measurements – which coincidently began in 1979, right after the Great Climate Shift of 1977.

Contrary to what you may have heard in news reports, the recent warming in the Arctic is probably not unprecedented.[24] It was just as warm in the late 1930s and early 1940s when the PDO was also in its positive, warm phase. There were newspaper reports of disappearing sea ice and changing wildlife patterns back then, too.[25] Most of the all-time high temperature records in the United States were set in the 1930s. The Northwest Passage

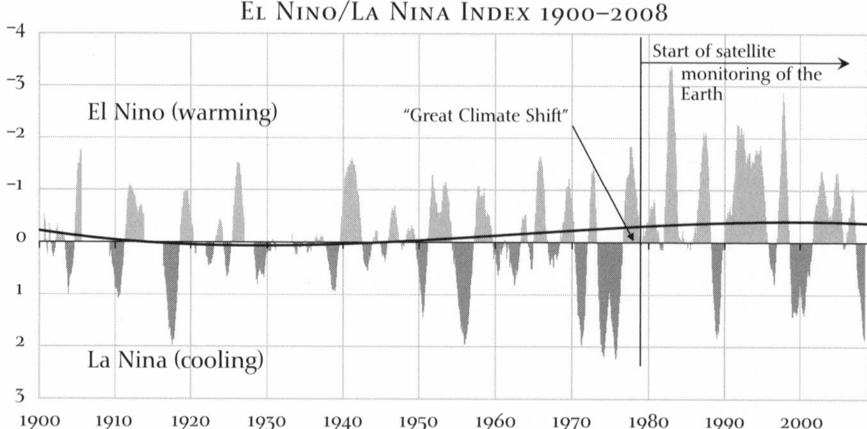

Fig. 5. Running five-month average values of the Southern Oscillation index for 1900–2008, which show more frequent El Niño activity since the 1980s.

was navigated without an icebreaker between 1940 and 1942, yet satellite observations of it opening up in 2007 were claimed to have recorded an unprecedented event.[26]

I am not the only one who believes that the correlation between these events and the PDO is more than just a coincidence. We will be learning much more about the potential role of the PDO in global warming in Chapter 6.

Examining the history of El Niño and La Niña over the twentieth century is also useful.[27] As can be seen in Fig. 5, these El Niño and La Niña events, collectively known as the Southern Oscillation, have occurred rather randomly throughout the twentieth century–at least until the 1980s. Since then, El Niños have been more frequent, as can be seen from the smooth curve fit to the data in Fig. 5. As mentioned previously, El Niño is known to cause anomalous global warmth, which raises the question of whether the Southern Oscillation is also part of what we call global warming. There is even a theory that the Southern Oscillation causes the PDO, and that we won't be able to forecast the PDO until we learn to forecast El Niño and La Niña.

Dr. Kevin Trenberth of the National Center for Atmospheric Research in Boulder, Colorado, has argued that manmade global warming has caused more El Niños in recent years. But I think it is much more likely that causation is actually operating in the opposite direction: more frequent El Niños might help provide a natural explanation for some of the warming observed in the late twentieth century. The general issue of cause-versus-effect is at the core of many mistakes that have been made in the interpretation of how the climate system works.

I cannot overemphasize that the potential importance of these and other modes of natural climate variability is their ability to change *global average cloudiness*. One of the primary mechanisms the Earth has for cooling itself is the production of clouds, which reflects some of the solar energy that reaches the Earth back to outer space. Because the average effect of clouds on the Earth's climate is to cool it, any natural change in global average cloudiness can also be expected to cause global warming or global cooling.

Natural Climate Variability Happens – With or Without the IPCC

The main theme of this chapter is that natural climate variability happens. All climate researchers agree that there is such a thing as short-term natural climate variability, yet you will almost never hear the terms "natural" and "global warming" in the same sentence. The IPCC has purposely avoided the issue. The most emphasis that the IPCC has ever placed on natural climate change has been with respect to the Medieval Warm Period and the Little Ice Age–and then it was only for the purpose of suppressing the historical evidence for their existence!

The IPCC's behavior in this regard is far from objective or scientific. There is no reason to assume that natural modes of climate change cannot occur on any time scale. As I pointed out earlier, the IPCC's view is not based on any in-depth investigation of the role of nature in climate change. They have merely found mankind alone at the scene of the crime, and so mankind must be responsible.

Fortunately, at least a few scientists associated with the IPCC appear to have decided that they can't ignore the fact that natural climate variability is a player in the global warming game. A paper published in 2008 claimed that natural climate variability associated with changing ocean currents might delay manmade global warming for at least another ten years.[28] And after a huge, coordinated protest at a coal-fired power plant in Washington D.C. was met by temperatures 25 deg. F below normal, a couple of climate experts admitted that unexpected natural cooling could delay "explosive" warming for up to thirty years.[29]

Why is it that even when ten-year forecasts of moderate global warming end up being wrong, we are still asked to believe in a forecast of explosive warming thirty years from now?

As noted previously, the 2007 report of the IPCC predicts an average warming trend for this century that is double what we have observed in the last thirty years. But rather than accelerate, warming seems to have stalled back in 2001. So, exactly when is the more rapid warming supposed to commence? The longer it takes to start, the more dramatic it will have to be in order for the IPCC's predictions to come true. Using an analogy from your everyday experience, the longer you wait to begin a car trip, the faster you will have to drive in order to get to your destination by a certain time.

Given the boldness of their predictions, if I were part of the IPCC leadership I would be sweating bullets by now. But then, given the short memory that the public has for long-term predictions from climatologists (or economists, or environmentalists), I think any long-term predictions of warming or cooling like this are pretty safe. It will be years, if not decades, before we know whether the IPCC was right or wrong. By that time, people will have been distracted by new climate forecasts. So, maybe the safest forecasts to make really are the ones that extend the furthest into the future. You will know tomorrow whether the forecast made by your favorite TV meteorologist is right or wrong, but no one will ever remember a thirty-year prediction made by a climate expert.

Unfortunately, the longest-range forecasts of climate change are also the ones that are most prone to exaggeration, and these are the forecasts that are now being used to guide policy. While climate forecasts provide some measure of entertainment for the public, they also carry with them a heavy burden of responsibility. Extremely important policy decisions involving the regulation of carbon dioxide emissions might be made in 2010, and these decisions will have to rely on the current state of the science. Laws passed by Congress or regulations issued by the EPA will necessarily be based on incomplete information and imperfect climate models.

So it is critical that we ask: How good are these climate models? What is the basis for their predictions of substantial – even catastrophic – levels of global warming? Can they be believed? Are climate models our high-tech prophets of the future, or are they just computer-generated Chicken Littles? Is the scientific consensus correct in saying that mankind now controls our climate, or is nature still in control?

In my first book, *Climate Confusion*, I provided some qualitative reasons why I thought the climate models are wrong. But we have learned much more since that book was published in early 2008. We have uncovered scientific evidence that strongly suggests the fears of manmade global warming are unfounded. While scientists like NASA's James Hansen and politicians like former Vice President Al Gore are increasingly warning us that we must act now to reduce carbon dioxide emissions, the latest science is turning in the opposite direction.

Even though global temperatures have not risen in at least seven years, we are being told that the climate is changing faster than expected. Even though Arctic sea ice extent recovered in 2008 and 2009 from its historic summer minimum in 2007, we are still being told that summer sea ice is melting faster than expected and might disappear entirely in the coming years. Contradictory claims are being made regularly. Yet these are the scientific warnings that the public and the policymakers have been relying on to form their opinions about global warming.

Some of the new research that I will be describing has been

published in the peer-reviewed scientific literature, but it is unlikely that you have heard about it. This is because the information gatekeepers – the news media – have refused to report on it. While the alarmists' publications get all the press, any research papers that cast doubt on the role of humanity in climate change are ignored. And while it is true that few articles on climate research ever argue directly against anthropogenic global warming, it is equally true that no published paper has ever ruled out natural causes for most of our warming. The few papers that have claimed to have ruled out natural sources of climate change are guilty of circular reasoning. They simply assume there are no natural sources of climate change, then run climate models built on that assumption, and end up showing that only mankind can explain global warming. Surprise, surprise.

The closest that researchers have come to ruling out natural causes of warming is to say something like, "We can't think of anything else that might have caused it," or "The extra carbon dioxide in the atmosphere is sufficient to cause the observed warming, so a natural mechanism of warming is not needed." The first statement is an admission of ignorance, the result of not bothering to look very hard for evidence of natural sources of climate change. I will return to that topic in Chapter 6.

The second statement rests on the belief that climate sensitivity is quite high, meaning that feedbacks in the climate system are positive – a belief that I will refute with satellite measurements in Chapter 5. While the climate sensitivity argument is subtle, it really is the single most important unresolved issue in global warming research. If we could measure the sensitivity of the climate system, we would know how much global warming will result from our addition of greenhouse gases to the atmosphere.

A few scientists are convinced that our climate is extremely sensitive, most notably NASA's James Hansen. The next chapter looks at just how extreme some of the warnings of climate catastrophe have been from those who believe in a very sensitive climate system.

Chapter 2 · "We Are Going to Destroy the Creation"

Worries over catastrophic global warming rest entirely on the belief that our climate system is very sensitive, that is, dominated by positive feedbacks, which amplify any warming or cooling influence. A few scientists are predicting planetary doom as a result of our burning of fossil fuels, and politicians are now using standard propaganda techniques to convince you that we must act quickly to save the Earth.

BEFORE I DEMONSTRATE how spectacularly wrong I think the scientific consensus is on global warming, I first want to illustrate how spectacular many of the global warming claims are in the first place. I find it paradoxical that while there has been no warming in recent years, and while public belief in a climate crisis is dwindling, the "experts" have been stepping up their rhetoric about the seriousness of the problem. This divergence cannot continue without someone eventually looking silly.

Back-to-back surveys released in January 2009 show that even as the scientific consensus on anthropogenic global warming is claimed to be getting stronger, the belief that global warming is just part of a natural cycle has become more widespread among the public. In the first survey, 82 percent of scientists polled responded positively to the question, "Has human activity been a significant factor in changing mean global temperatures?"[1] Meanwhile, *Rasmussen Reports* found in a national phone survey that 44 percent of Americans believe global warming to be part of a natural cycle, whereas only nine months earlier the figure was 34 percent.[2] An annual Pew survey on the importance of twenty domestic issues to American citizens found that global warming ranked dead last.[3] Apparently, Al Gore's $300 million advertising blitz and T. Boone Pickens selling wind energy on TV have not had the intended effect on public opinion.

This makes it all the more bizarre when a few scientists out on the fringe of global warming research claim that we are in the middle of a climate crisis. The most vocal scientist on the subject has been NASA's James Hansen, an astrophysicist by training who has turned himself into both a climate modeler and a developer of one of the leading global temperature datasets. And he has not been shy about expressing his opinions on the subject.

On September 23, 2008, Hansen told an audience of several hundred in Topeka, Kansas, that if we don't get our emissions of carbon dioxide under control, "we are going to destroy the creation."[4] On June 23, 2008, the twentieth anniversary of his famous 1988 testimony before Congress for then Senator Al Gore, he wrote in a *Guardian* editorial, "Fossil fuel companies know what the story is. I think they're guilty of crimes against humanity and nature."

In testimony before the Iowa Utilities Board on October 22, 2007, Dr. Hansen told of driving past a long train loaded up with coal. "If we cannot stop the building of more coal-fired power plants," he said, "those coal trains will be death trains – no less gruesome than if they were boxcars headed to crematoria, loaded with uncountable irreplaceable species."[5] Hansen later apologized for this statement after complaints that his analogy had trivialized the real Holocaust.

In January 2009, on the eve of President Obama's inauguration, Hansen declared, "We cannot afford to put off change any longer. We have to get on a new path within this new administration. We have only four years left for Obama to set an example to the rest of the world. America must take the lead."[6]

These are bold statements indeed. Dr. Hansen, like Al Gore, has called for civil disobedience to block the construction of any more coal-fired power plants. Of course, if Hansen is correct and we continue business as usual with our burning of coal and petroleum, there won't be a habitable world in which to conduct business, will there? If the climate system is as sensitive as Dr. Hansen believes, we are indeed in trouble, and no cost is too high to save both ourselves and the rest of the creation.

But if he is wrong, we run the risk of killing literally millions of people. The risk that Hansen and Gore warn about is theoretical, based on calculations in computer models. In contrast, history has taught us that the risks associated with poverty are very real. Penalizing the use of our most inexpensive energy sources will destroy wealth and will lead to starvation for many of the world's malnourished. Millions more will become susceptible to food-borne illness if they can no longer afford a refrigerator.

And what if more carbon dioxide in the atmosphere is actually a good thing for life on Earth? I will address that possibility in Chapter 7.

While many applaud Gore's and Hansen's "courage" in speaking out on global warming, I can't help but compare their rhetoric to yelling "Fire!" in a crowded theater. If there is no fire, people may be needlessly killed in the resulting rush for the exits. Yet these alarmists continue to be lauded and showered with awards.

I do not really care where our energy comes from. If we can develop relatively clean alternative energy sources that are cost-competitive, then fine. But I do care that we have energy and that as many people have access to it as possible. Since nearly one billion of the world's poor do not even have electricity, denying them the chance to acquire it is, in my view, immoral. Environmentalists who do not live in Africa or India have blocked the construction of hydroelectric dams there, depriving the poor of access to the electricity that we take for granted.[7] Roy Innis, a civil rights veteran and leader of the Congress of Racial Equality, has been particularly outspoken on this subject. As he explains in his book *Energy Keepers, Energy Killers: The New Civil Rights Battle*, bad policies based on bad science will disproportionately hurt the poor and minorities.[8]

Poverty is a very real threat, and it exists today. The climate alarmists' threat is theoretical, and it concerns events that have not yet happened and probably never will. There is no convincing evidence that anyone has ever been killed by manmade global warming. But we have daily evidence that poverty kills.

If Dr. Hansen is correct and humans are responsible for the recent warming, then what caused earlier periods of dramatic warming–and cooling? Has natural climate change now ended, having been replaced by human-caused climate change? This seems unlikely.

THE VOSTOK ICE CORE RECORD

Why does James Hansen believe that global warming is so strongly driven by carbon dioxide? Rather than relying on real measurements of the climate system today, he mostly depends on his interpretation of ice core reconstructions of temperature and CO_2 variations over hundreds of thousands of years from the Vostok ice core recovered from the Antarctic ice sheet.

This evidence was showcased in Al Gore's movie, *An Inconvenient Truth*. In fact, I thought it was probably the most effective part of that movie–at least to anyone who does not know the details that Mr. Gore left out. He used a chart like that in Fig. 6 as a huge stage prop to demonstrate how CO_2 and temperature have gone up and down together, apparently in lockstep, over the last half-million years. At the very end of the graph, Gore showed how the CO_2 content of the atmosphere has recently skyrocketed, presumably due to our burning of fossil fuels. To emphasize the point, Mr. Gore even used a man-lift to reach the top of the graph.

The implication was clear: atmospheric CO_2 and temperature have always gone up and down together, and now the atmospheric concentration of CO_2 is so high that we are surely in for catastrophic levels of global warming. From what I can tell, this is the main source of Hansen's alarmist views. The atmospheric CO_2 content is now well above what we think it has been over the last million years, a period during which carbon dioxide and temper-

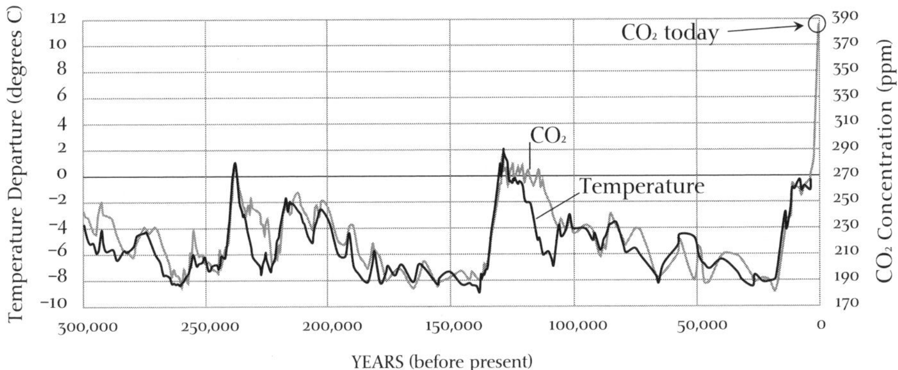

Fig. 6. Estimates of atmospheric temperature and carbon dioxide concentration over the last 300,000 years from the Vostok ice core in Antarctica. This is similar to a stage prop used by Al Gore in An Inconvenient Truth.

ature seem to have gone up and down in lockstep, and so we must surely be in for catastrophic warming as carbon dioxide concentrations continue their upward march.

A truth that was conveniently left out of Mr. Gore's presentation is related to an issue that I will emphasize throughout this book: *cause versus effect*. Mr. Gore, whether he knew it or not, was assuming that the hundreds of thousands of years of CO_2 variations were causing the temperature variations – and not the other way around. But as the climate researchers who produce the Vostok dataset well know, there is an average 800-year lag between these two variables, with the temperature changes preceding the CO_2 changes.[9] At face value, this would suggest that the temperature changes caused the CO_2 changes. I will discuss temperature causing atmospheric CO_2 changes more in Chapter 7.

To be fair, Hansen is well aware of this time lag, which appears to argue against the claim that more carbon dioxide causes warming. But he claims that temperature changes causing CO_2 changes in the ice core record just got the ball rolling, and then the direction of causation magically switched to CO_2 forcing temperature change.

The argument goes something like this: There are long-term

cycles, called Milankovitch cycles, in the Earth's tilt and orbit around the sun. These cycles cause small fluctuations in how much sunlight reaches different parts of the Earth. The prevailing opinion is that the resulting variations in sunlight are not strong enough to have caused the ice age cycles shown in Fig. 6 unless there are positive feedbacks amplifying that small amount of forcing. That is, unless the climate system is very sensitive. When the Milankovitch cycles cause a small amount of warming, it leads to an increase in the CO_2 content of the atmosphere. But since more CO_2 also causes warming, this sets up a vicious cycle of warming, then more CO_2, then more warming. The process supposedly reverses when the Milankovitch cycles switch to causing a small decrease in the sunlight reaching the Earth. A vicious cycle then occurs in the opposite direction, with decreasing CO_2 and falling temperatures plunging the Earth into an ice age.

This whole argument is speculative, at best. Is it theoretically possible? I suppose so. In the climate business, as in any other realm of scientific investigation, there is always some way to make the evidence fit your theory.

But if the major forcing of temperature really is carbon dioxide, as Hansen claims, then the observed time lag either should be reversed or should not be there at all. Therefore, the fact that the temperature changes preceded the CO_2 changes in the ice core record is, to me, sufficient evidence that CO_2 was not the forcing of, but instead the response to, the temperature changes.

I have to wonder: what if the CO_2 changes actually preceded the temperature changes in the ice core record? I'm quite sure that Hansen would have accommodated it into his theory, since it would be the obvious expectation for causation in that direction.

But the biggest objection to the theory that the Milankovitch cycles caused the ice ages is that there is no statistically significant connection between the two! A careful analysis has shown that the timing of the Milankovitch cycles relative to the ice ages is no closer than what would be expected by chance.[10]

I believe that the ice core record is largely irrelevant to what is happening today. There are so many things we don't know about how the climate system operated hundreds of thousands of years ago. Geologists point to periods millions of years ago when atmospheric carbon dioxide concentrations were much higher than they are today, or episodes of great warmth in the Arctic, as evidence of natural climate change. But I don't believe we have a clue what the governing factors were for these events. As it is, our best Earth-observing satellites covering the globe every day are providing information that leads various scientists to different conclusions. How can we hope to know what, if anything, the conditions on Earth in the distant past have to do with how the climate system operates today? Are we going to make policy decisions that cause immense human suffering on the basis of a speculative theory about how global temperatures might have responded to a forcing, when there might not even be a statistically significant relationship between the two anyway? I sure hope not.

Global Warming Propaganda

The methods used by global warming alarmists to convince you that more carbon dioxide is going to ruin the Earth are increasingly laced with insults and attacks on anyone who might disagree with them. At the Live Earth concert at Giants Stadium on July 7, 2007, Robert F. Kennedy Jr., an environmental lawyer, shouted himself hoarse as he called those who are skeptical of mankind's role in global warming "traitors" and "corporate toadies."

Al Gore's tactics have been a little more subtle, and reminiscent of propaganda methods that have proved throughout history to be effective at influencing public opinion. Listed below are fifteen propaganda techniques I have excerpted, sometimes paraphrased, from a Wikipedia page on the subject. Beneath each are quotations from Mr. Gore as he has attempted to goad the rest of us into reducing our CO_2 emissions. Except where otherwise indicated, the quotations come from his testimony before the U.S. Senate Environment and Public Works Committee on March 21, 2007. I'll let you decide whether the shoe fits.

APPEAL TO FEAR: Appeals to fear seek to build support by instilling anxieties and panic in the general population.

> "I want to testify today about what I believe is a planetary emergency – a crisis that threatens the survival of our civilization and the habitability of the Earth."

APPEAL TO AUTHORITY: Appeals to authority cite prominent figures to support a position, idea, argument, or course of action. Also, TESTIMONIAL: Testimonials are quotations, in or out of context, especially cited to support or reject a given policy, action, program, or personality. The reputation or the role (expert, respected public figure, etc.) of the individual giving the statement is exploited.

> "Just six weeks ago, the scientific community, in its strongest statement to date, confirmed that the evidence of warming is unequivocal. Global warming is real and human activity is the main cause."

> "The scientists are virtually screaming from the rooftops now. The debate is over! There's no longer any debate in the scientific community about this." (from *An Inconvenient Truth*)

BANDWAGON: Bandwagon and "inevitable-victory" appeals attempt to persuade the target audience to join in and take the course of action that "everyone else is taking." Also, JOIN THE CROWD: This technique reinforces people's natural desire to be on the winning side. This technique is used to convince the audience that a program is an expression of an irresistible mass movement and that it is in their best interest to join.

> "Today, I am here to deliver more than a half million messages to Congress asking for real action on global warming. More than 420 Mayors have now adopted Kyoto-style

> commitments in their cities and have urged strong federal action. The evangelical and faith communities have begun to take the lead, calling for measures to protect God's creation. The State of California, under a Republican Governor and a Democratic legislature, passed strong, economy wide legislation mandating cuts in carbon dioxide. Twenty-two states and the District of Columbia have passed renewable energy standards for the electricity sector."

FLAG-WAVING: An attempt to justify an action on the grounds that doing so will make one more patriotic, or in some way benefit a group, country, or idea. Also, INEVITABLE VICTORY: Invites those not already on the bandwagon to join those already on the road to certain victory. Those already or at least partially on the bandwagon are reassured that staying aboard is their best course of action.

> "After all, we have taken on problems of this scope before. When England and then America and our allies rose to meet the threat of global Fascism, together we won two wars simultaneously in Europe and the Pacific."

AD HOMINEM: A Latin phrase which has come to mean attacking your opponent, as opposed to attacking their arguments. Also, DEMONIZING THE ENEMY: Making individuals from the opposing nation, from a different ethnic group, or those who support the opposing viewpoint appear to be subhuman.

> "You know, 15 percent of people believe the moon landing was staged on some movie lot and a somewhat smaller number still believe the Earth is flat. They get together on Saturday night and party with the global-warming deniers." (October 24, 2006, Seattle University)

(This one is especially humorous now that two astronauts who have actually walked on the moon, Harrison Schmitt and Buzz

Aldrin, have voiced their skepticism about global warming being the fault of humans.)

Appeal to prejudice: Using loaded or emotive terms to attach value or moral goodness to believing the proposition.

> "And to solve this crisis we can develop a shared sense of moral purpose." (June 21, 2006, London, England)

Black-and-white fallacy: Presenting only two choices, with the product or idea being propagated as the better choice.

> "It is not a question of left vs. right; it is a question of right vs. wrong." (July 1, 2007, *New York Times* op-ed)

Euphoria: The use of an event that generates euphoria or happiness, or using an appealing event to boost morale.

> Live Earth concerts organized worldwide in 2007 by Al Gore.

Disinformation: The creation or deletion of information from public records, for the purpose of making a false record of an event or the actions of a person or organization. Pseudo-sciences are often used to falsify information.

> "Nobody is interested in solutions if they don't think there's a problem. Given that starting point, I believe it is appropriate to have an over-representation of factual presentations on how dangerous (global warming) is, as a predicate for opening up the audience to listen to what the solutions are, and how hopeful it is that we are going to solve this crisis." (May 9, 2006, *Grist* interview) Also, the widely publicized "Climategate" release of email correspondence between leading IPCC scientists in late 2009 revealed plans to hide or destroy temperature data not supporting

the IPCC's efforts, as well as manipulation of the peer review process.

Stereotyping or name calling or labeling: This technique attempts to arouse prejudices in an audience by labeling the object of the propaganda campaign as something the target audience fears, hates, loathes, or finds undesirable. Also, Obtain disapproval: This technique is used to persuade a target audience to disapprove of an action or idea by suggesting that the idea is popular with groups hated, feared, or held in contempt by the target audience.

> "There are many who still do not believe that global warming is a problem at all. And it's no wonder: because they are the targets of a massive and well-organized campaign of disinformation lavishly funded by polluters who are determined to prevent any action to reduce the greenhouse gas emissions that cause global warming out of a fear that their profits might be affected if they had to stop dumping so much pollution into the atmosphere." (January 15, 2004, New York City)

It is unfortunate that such tactics are used to push an agenda that is driven more by quasi-religious beliefs and financial and political motives than by an objective assessment of the science. Science itself is being misused to advance policy goals that would never be embraced on their own merits.

Of course, Al Gore, James Hansen, and Robert F. Kennedy Jr. are not the only activists who have been sounding the global warming alarm. A gaggle of entertainment personalities – George Clooney, Sheryl Crow, Julia Roberts, and Leonardo DiCaprio, for instance – have decided they must help inform the teeming masses that we are consuming too much. The hypocrisy of such claims by these most voracious consumers of energy and natural resources is evident to everyone. Everyone except the celebrities, that is.

The news media have also been complicit in this campaign to misinform the public. As I said earlier, their silence on any published science that runs counter to Gore & Company is one of the reasons I wrote this book. While some have likened global warming skeptics' scientific research to Big Tobacco–funded research that supposedly showed smoking was not dangerous, I would say the media's refusal to report on skeptics' peer-reviewed research is like the tobacco company executives' suppression of evidence. Some may claim that the media's silence is a way of trying to help people–that the tobacco executives were trying to save their own butts, so to speak. But this would mean that the media are putting themselves in the position of deciding what is good for people, rather than just keeping us informed.

I'm sorry, but from what I've seen of the curricula in journalism schools, I would rather have our elected representatives, economists, or even scientists handling those decisions, not journalists.

It's not that Mr. Gore's views lack any basis in science whatsoever. A few scientists like James Hansen appear to be genuinely worried about humanity's future in a warming world. But their predictions for the future–and their explanations for climate change in the past–are totally dependent on the existence of high climate sensitivity, meaning positive feedbacks in the climate system. And observational evidence that today's climate system behaves in this way is almost nonexistent.

I will present new evidence for an insensitive climate system, one dominated by negative feedbacks. This means that Earth's climate will not change nearly as much in response to our carbon dioxide emissions as is widely claimed. But first, since "effect" is preceded by "cause," a better understanding of feedback needs to be preceded by a discussion of forcing.

Chapter 3 · Forcing: How Warming Gets Started

Whether it is the Earth's climate, or a pot of water on the stove, a temperature change is always caused by an imbalance between energy gained and energy lost.

The temperature of a pot of water will change as long as there is an imbalance between the heat gained from the stove and the heat lost by the pot to its surroundings. In the case of the Earth, global warming happens from an energy imbalance caused by either (1) an increase in absorbed solar radiation, or (2) a decrease in thermally emitted infrared radiation lost to outer space. Modern global warming theory starts with manmade greenhouse gases causing a small reduction in the Earth's ability to cool to outer space.

THE MODERN THEORY of global climate change involves two main components: *forcing* and *feedback*. They are not difficult to grasp because they are routinely experienced by everyone. Understanding these concepts will provide the insight you need to penetrate the mysterious veil surrounding the IPCC's climate models and their forecasts of dangerous levels of warming.

If you understand why the inside of a car sitting in the sun heats up so much, you can understand forcing. It is what initiates a temperature change. And if you understand how rolling down the car's window reduces the warming, you can then understand feedback. It determines how large the temperature change will be in response to the forcing. As we will see, it's in the realm of feedbacks that the IPCC has made some fundamental errors. Later, in Chapter 6, I will be using a simple climate model based on forcing and feedback – one you can run on your home computer – to demonstrate how global warming can arise from what the climate system does naturally.

THE FORCING OF TEMPERATURE CHANGE: ENERGY IMBALANCE

There is a common misconception about the reason why the temperature of anything changes. Here is a question:

TRUE OR FALSE? The temperature of a pot of water placed on a hot stove is determined by how much heat is transferred from the stove to the pot.

ANSWER: False.

More precisely, the statement is only half true. With the help of Fig. 7, let's examine what happens when we place a pot of water at room temperature on a hot stove. We will assume it is a gas stove so that the forcing (the flame) is always on, rather than

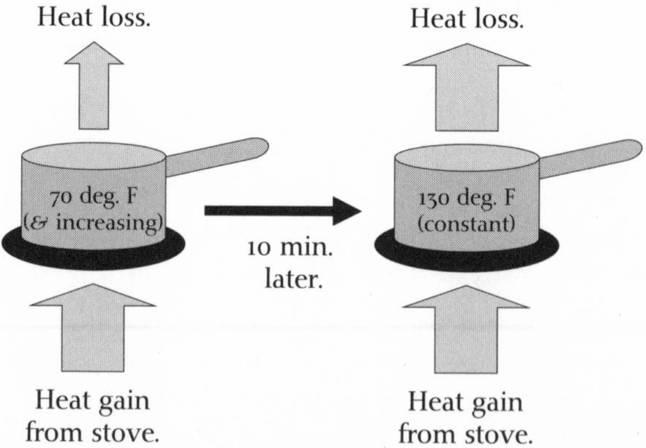

Fig. 7. The temperature of a pot of water placed on a hot stove will increase until the rate of heat loss by the pot equals the rate of heat gain from the stove.

cycling on and off as an electric stove does. Also, we will set the flame on low so that the water does not boil.

Heat begins to flow into the pot of water, and the temperature of the water starts to rise. It gets warmer and warmer until at some point, maybe ten minutes later, the temperature of the water stops increasing. It reaches a certain temperature and then stays there, achieving what scientists call a steady state, or equilibrium.

But how can the water stop warming if the stove's flame is still transferring heat into the pot? Clearly, the temperature change of the water is more than just a matter of adding heat generated by the stove; otherwise the temperature would keep rising as long as the stove was on.

What we are missing is the fact that the pot is losing heat to its surroundings at the same time it is gaining heat from the stove. What controls whether the temperature of the water goes up or down is not whether the stove is pumping energy into it, but whether there is an imbalance between (1) the energy *gained* by the pot from the stove and (2) the energy *lost* by the pot to its surroundings. This concept is called "energy balance," and it is fundamental to understanding why the temperature of anything goes up or down – including the Earth.

So let's examine what really happens when you heat a pot of water. As soon as the water starts to warm up, it begins to evaporate more rapidly, thus losing latent (stored) heat to the air. At the same time, convective air currents in direct contact with the pot take away heat. Third, infrared radiation emitted by the warm pot to its cooler surroundings causes heat loss. Infrared radiation is the radiant heat you feel at a distance from hot objects, for instance from a roaring fire or a brick wall that has been warmed by the sun all day. Virtually everything gives off infrared radiation, at any temperature, and the hotter something is, the more infrared energy it emits.[1]

As the pot becomes warmer and warmer, these three heat-loss mechanisms speed up. The water evaporates faster, the warm convective air currents increase, and the loss of infrared energy by the pot to its surroundings intensifies. Finally, a temperature is reached at which the rate of heat loss by the pot to its surroundings equals the rate of heat gain by the pot from the stove. It is only at this point of energy balance that the temperature stops changing.

Now we can better understand what happens when the temperature of something changes: *The temperature of an object will increase as long as the rate of heat gain exceeds the rate of heat loss by the object.*

The reverse is also true: *The temperature of an object will decrease as long as the rate of heat loss exceeds the rate of heat gain by the object.*

As long as there is an imbalance of energy flows in and out of an object, its temperature will change. Once energy balance is reached, the temperature stops changing. It's as simple as that. And this is exactly the same concept that is involved in global warming theory.

These rules of temperature change have only one exception: phase changes between the liquid, solid, and gaseous states of matter. In the case of water, one phase change occurs at 0 deg. C (32 deg. F), water's freezing and melting point. It takes extra heat to change ice into liquid water, and that extra heat is given up when liquid water freezes to form ice. Similarly, evaporation of water (change from the liquid phase to vapor phase) requires extra heat, which is released when water vapor condenses back into liquid water (in steam or a cloud).

Aside from these specific cases of a change in phase, we can use "energy imbalance causes a temperature change" as a good rule of thumb.

The concept of energy balance explains all kinds of everyday experiences that we normally do not think about. For instance, after you start your car the engine begins to warm up as heat from the burning fuel accumulates within it. But as the engine gets warmer, it loses heat to its surroundings at an increasing rate. The hotter the engine gets, the faster that heat is lost. The temperature of the engine finally stops increasing when the rate of heat loss by the engine to its surroundings equals the rate of heat gain from burning fuel. In order to make sure the engine operates at maximum efficiency, a thermostat inside the engine forces that energy balance to occur at a specific temperature, adjusting the flow rate of coolant through the engine so that the engine's temperature remains close to 190 or 195 deg. F.

We can also use the example of heating your home in the winter. In order to keep the inside of your home at a constant temperature, the heating system must generate heat at a rate that is equal to the rate of heat loss by the house to the outside. If your house is well insulated, the rate of heat loss will be slow, and so the heating system will not have to work as hard. If you want the temperature to be higher, you tell the heating system (through the thermostat setting) to transfer heat into the house at a faster rate. This causes an energy imbalance, so the temperature in the house increases. But it also increases the rate at which heat is lost by the house to the outside. The temperature stops going up when energy balance is once again restored, and the thermostat then turns the heat off.

Now let's apply the concept of energy balance to global warming. We will replace the pot of water with the Earth, and the stove with the sun (Fig. 8). While the sunlit side of the Earth absorbs solar energy, both the day and night sides of the Earth are continuously losing energy to outer space through infrared radiation.

Note that unlike the pot of water, which loses heat through three mechanisms, the Earth can lose heat to outer space through only one mechanism: infrared radiation. There is no air in outer space to carry heat or moisture away, so the Earth can cool only through the "heat radiation" it loses to the cold depths of outer space.

The energy balance of the Earth is therefore, in some sense, simpler than the energy balance of a pot of water on the stove. It is determined by only two energy flows: the rate at which solar energy is absorbed, and the rate at which infrared energy is lost to outer space. These are the two flows of radiant energy that are involved in global warming or global cooling: absorbed sunlight and emitted infrared radiation. If you are wondering about the flow of heat from the Earth's core to the surface, that flow is very

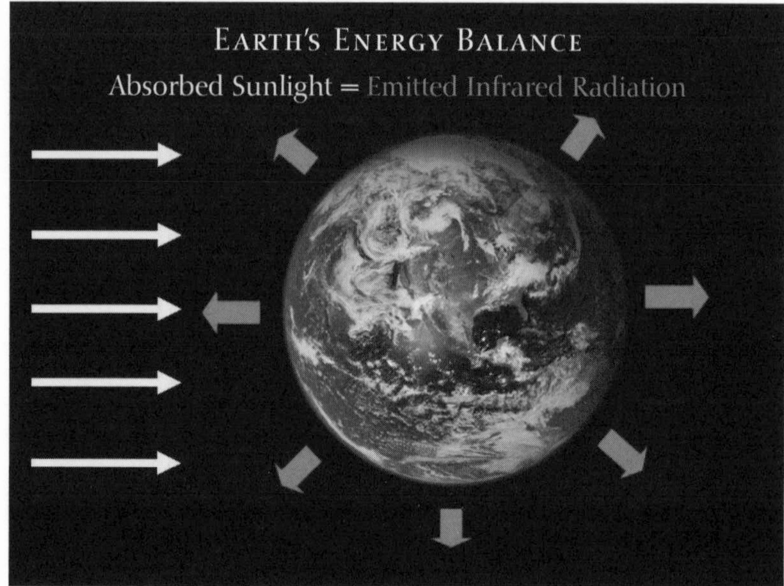

Fig. 8. The temperature of the Earth remains constant as long as the rate of energy input by sunlight equals the rate of energy loss by infrared radiation.

weak, estimated to be a small fraction of 1 percent of the average rate of solar heating.

The scientific consensus is that these rates of energy flow in and out of the Earth have remained the same and in balance for centuries – if not millennia – and that the Earth's temperature has therefore remained the same. Since only radiant flows of energy are involved, we call this kind of energy balance "radiative energy balance," or just radiative balance. But if something causes a radiative imbalance, we call this "radiative forcing." If some forcing agent causes the amount of absorbed sunlight to be unequal to the amount of infrared radiation being emitted to outer space, there is radiative forcing present, and the average temperature of the Earth can be expected to change.

It cannot be overemphasized that this concept of radiative energy balance is central to understanding global warming – or global cooling. It is the starting point. In the theory of manmade global warming, the Earth is assumed to have been in a state of energy balance before mankind came along and upset that balance. Humans have knocked it out of balance by adding greenhouse gases, thus reducing the Earth's ability to cool to outer space.

The very first figure in this book, however, showed that this assumption of a long-term energy balance of the Earth cannot be supported. The average temperature of the climate system has probably never stayed constant, which means that the flows of energy in and out of the climate system must have been changing, too.

Now we can better understand what must happen for global warming or global cooling to occur:

Global warming will occur if the amount of sunlight absorbed by the Earth is increased (e.g. from less low cloud cover), or if the amount of infrared radiation lost to space is decreased (e.g. from more greenhouse gases, more water vapor, or more high cloud cover).

Global cooling will occur if the amount of sunlight absorbed by the Earth is decreased (e.g. from more low cloud cover), or if the amount of infrared radiation lost to space is increased (e.g. from less greenhouse gases, less water vapor, or less high cloud cover).

These changes do not have to occur over the whole Earth, just over regions large enough either to substantially affect the global average, or to cause global weather systems to spread the influence of these regional changes to cover other regions as well.

Next I will explain the greenhouse gas theory of global warming.

Global Energy Imbalance Caused by Increasing CO_2

Judging from media reports and the comments of some scientists and politicians, you would think that the main role of carbon dioxide was to pollute the atmosphere and scorch the Earth. Nothing could be further from the truth. Carbon dioxide is necessary for life on Earth. Photosynthesis by plants and phytoplankton in the ocean (the start of the food chain there) would be impossible without CO_2; and without photosynthesis, animals and humans would be done for, too.

Given the necessity of carbon dioxide in the atmosphere for life on Earth, it is surprising that the atmosphere currently contains 540 times as much oxygen as it does carbon dioxide. The atmosphere of Venus has over 230,000 times as much carbon dioxide as does our atmosphere. Even Mars, which has only about 1 percent of the atmospheric density as Earth, still has twelve times as much CO_2 in absolute terms. As will be discussed more in Chapter 7, a natural conclusion one could reach from these facts is that life on Earth is sucking as much carbon dioxide as it can out of the atmosphere.

But carbon dioxide also absorbs and emits infrared energy, which makes it a so-called greenhouse gas. Greenhouse gases act like a radiative blanket within the atmosphere, warming the lower atmosphere and cooling the upper atmosphere. By far the most important greenhouse gas in the atmosphere is water vapor. Clouds also have a strong greenhouse effect, especially relatively thin high-altitude clouds. Water vapor and clouds account for about 90 percent of the Earth's natural greenhouse effect, CO_2 amounts to about 3.5 percent, and methane contributes even less.[2]

The consensus explanation of manmade global warming says that our greenhouse gas emissions – mainly carbon dioxide – have caused a slight increase in the greenhouse effect. By itself, the accumulated energy imbalance that has been building up for over 100 years is small, amounting to about 0.6 percent of the radiant flows of energy in and out of the Earth. Yet it is claimed that, since most of the extra carbon dioxide in the atmosphere stays around for decades, the persistence of the energy imbalance will result in long-term warming of the global atmosphere. So, even though it is a relatively minor greenhouse gas, adding more CO_2 to the atmosphere can be expected to cause some level of warming. The big question is, how much?

By way of comparison, a major volcanic eruption like Mt. Pinatubo in the Philippines in 1991 can cause a relatively large energy imbalance, as much as 2 percent, in a matter of months.[3] Fortunately, the sulfuric acid aerosols in the stratosphere that cause this imbalance persist for only a couple of years, and so their cooling effect has little long-term consequence for the climate system.

How do we know that the CO_2 content of the atmosphere has been increasing? Accurate measurements of CO_2 have been made at Mauna Loa, Hawaii, since 1958 (see Fig. 9), and at a number of other monitoring stations around the globe starting in more recent years.[4] The measurements at Mauna Loa are automatically made and recorded on a continuous basis throughout the day, and the data are quality-controlled so that local influences such as volcanic vents and agricultural fields are minimized.

The tiny yearly squiggles in Fig. 9 are believed to result from the seasonal inhaling and exhaling of carbon dioxide by vegetation growing and dying each year in the Northern Hemisphere. Growing vegetation requires CO_2 for photosynthesis, which removes more CO_2 from the atmosphere and causes the downward parts of the wiggles. In the winter, much of that new foliage dies and decays, giving off extra CO_2 and causing the upward part of the wiggles.

In the context of global warming, it is not the yearly wiggles we are interested in, but the long-term upward trend. As of 2008, the atmospheric CO_2 concentration was approaching 390 parts per million (ppm) by volume, or about 39 molecules of CO_2 per 100,000 molecules of air. This is close to 40 percent higher than the concentration that is believed to have existed before the industrial revolution, 270 ppm.

Al Gore likes to dramatize our CO_2 emissions by saying that mankind pumps about 70 million tons of CO_2 into the atmosphere every day. But this statistic is put into perspective when one realizes that it takes five years of those daily greenhouse gas emissions to add just one molecule more of CO_2 to every 100,000 molecules of air in the atmosphere. Thus, even though carbon dioxide is a "trace" gas (meaning there isn't much of it in the atmosphere), it still takes a long time for our burning of fossil fuels to impact its atmospheric concentration substantially. Since the global atmosphere is a pretty big place, 70 million tons of CO_2 a day is a vanishingly small amount.

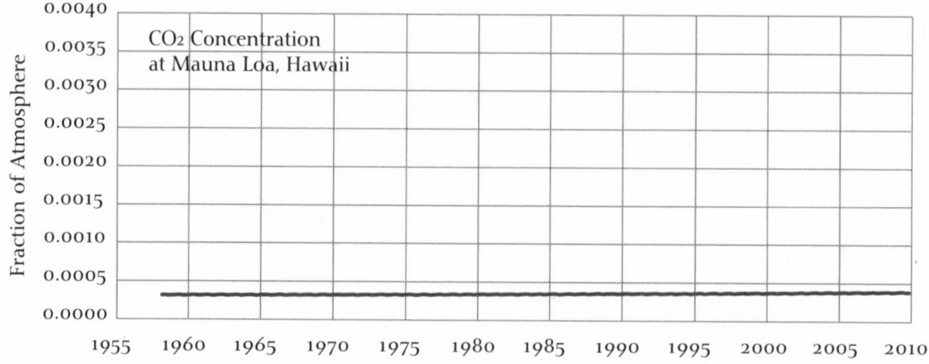

Fig. 9. CO_2 concentrations at Mauna Loa, Hawaii, from 1958 to 2008.

To emphasize just how small the current atmospheric concentration of carbon dioxide is, you would have to extend the top of the graph in Fig. 9 upward by about 400 feet before you reached 100 percent of the atmosphere. For even more fun, let's return to Al Gore's stage prop in *An Inconvenient Truth* (Fig. 6 in Chapter 2), where he used a man-lift to exaggerate how high the CO_2 content of the atmosphere has become. In order to reach the top of the graph, representing 100 percent of the atmosphere, that man-lift would have to raise him nearly one mile upward.

The point is that there is very little CO_2 in Earth's atmosphere, and it takes a very long time for humanity to impact it by a measurable amount.

Mr. Gore also likes to compare the Earth's greenhouse effect to that on Venus, where surface temperatures are hot enough to melt lead. Is the high surface temperature on Venus mostly the result of its greenhouse effect? Yes. Is this relevant to Earth and global warming? No. It will have taken humanity hundreds of years of greenhouse gas emissions to increase the CO_2 content of the Earth's atmosphere by a factor of two, and remember that the CO_2 concentration in the atmosphere of Venus exceeds that of Earth by a factor of 230,000.

According to global warming theory, the extra carbon dioxide causes a small forcing, which is to say a small warming tendency. As it warms, the Earth then emits more infrared radiation to space (because a warmer object always emits more infrared radiation than a cooler one) until a new state of energy balance is reached. And just as the pot of water heating on the stove eventually stops warming when its rate of energy loss finally equals its rate of energy gain, the Earth's temperature stops rising when the radiant energy lost to space once again equals the amount absorbed from the sun, and energy balance is restored. Qualitatively, this constitutes the basic explanation of what happens in anthropogenic global warming.

The Direct Warming Effect of Doubling CO_2

Now for some quantitative details. Sometime late in this century, the CO_2 concentration in the atmosphere will probably be at least double the estimated preindustrial level: 540 ppm compared with around 270 ppm. This doubling of carbon dioxide has been dubbed "$2 \times CO_2$," and it has provided a useful milestone for comparing different estimates of future warming.

How much warmer would the Earth become as a result of this doubling of CO_2, if nothing else in the climate system changed except the temperature? We cannot answer this question experimentally because we cannot put 100 miles of the vertical depth of the atmosphere in the laboratory, with its wide range of air pressures and temperatures. The answer can only be estimated with theoretical computations, usually using a one-dimensional (vertical only) radiative-convective model run on a computer.

The computations are not simple. They require quantitative estimates of how air absorbs infrared radiation at different wavelengths, temperatures, and air pressures. This information is then put into a computerized radiative transfer model that calculates the flows of infrared energy up and down at many levels throughout the depth of the atmosphere, at the same time that the sun is heating the surface, and atmospheric convection is transporting much of that heat energy from the surface to higher in the atmosphere. For global average conditions, these models do a pretty good job of reproducing the observed vertical profile of temperature in the atmosphere. In fact, the answers we get out of such models haven't changed much in the last forty years, even as the models have been improved.

When the models are run with $2 \times CO_2$, they produce a global average reduction in the rate of infrared cooling of the Earth by about 3.7 watts per square meter, a value which I will assume to be correct through the rest of this book. But what you might be surprised to learn is that after the model reaches a new state of energy equilibrium, the direct warming effect of $2 \times CO_2$ is *only about 1 deg. C (1.8 deg. F)*. And since atmospheric convection typically causes more warming at high altitudes than near the surface, the surface warming can amount to only 0.5 deg. C (about 1 deg. F).

Let's summarize: If we assumed that the Earth was initially in energy balance, and then instantly doubled the atmospheric concentration of CO_2, there would then be an energy imbalance of 3.7 watts per square meter. This means that the rate at which infrared energy is lost by the Earth to space would suddenly be 3.7 watts per square meter less than the amount of sunlight being absorbed by the Earth. This radiative forcing would then cause the Earth to warm by about 1 deg. C over a period of years before enough extra infrared radiation was emitted to space to restore energy balance.

The value of 1 deg. C is the estimate of global warming *without* feedbacks; only the temperature has been assumed to change, and nothing else. This value can only be computed theoretically: there is no laboratory experiment we can perform to actually prove that it happens. Our satellite instruments still do not have the absolute accuracy to measure the small imbalance from Earth orbit that is believed to exist from more carbon dioxide in the atmosphere, so we cannot even directly measure the mechanism that supposedly causes global warming!

As of 2009, it is estimated that humanity's CO_2 emissions have increased the atmospheric CO_2 concentration by close to 40 percent. This has caused an estimated 1.6 watts per square meter of extra energy to be trapped, out of the estimated 235 to 240 watts per square meter that the Earth on average emits to outer space on a continuous basis. We really don't know the exact magnitude of the average flows of energy in and out of the Earth to better than several watts per square meter. It could be 235, 240, or 245 watts per square meter.

I find it amazing that the scientific community's purported near-certainty that global warming is manmade rests on a forcing mechanism – a radiative imbalance – that is *too small to measure.* In contrast to this theoretical forcing, I will later describe a natural global warming mechanism that can be measured from our satellites.

The Character of the Temperature Response to an Energy Imbalance

We have established that an energy imbalance is what causes a temperature change – of the Earth or of a pot of water on the stove. What does that temperature response to forcing look like over time? In Fig. 7 we saw that an energy imbalance imposed on a pot of water by a hot stove will cause the temperature to rise until the energy imbalance is relieved, but we didn't discuss what that transition looks like.

Whether it is the Earth warming up, or the pot of water on the stove, or a car sitting in the sun, the resulting change in temperature with time looks basically the same; the only difference is the total amount of time involved. If we were to make temperature measurements of the pot of water from the time we placed the pot on the stove, until the time that the water stopped warming up (reaching energy balance), the data would look something like what I have plotted in the top panel of Fig. 10.

After heat is applied to the pot of water, the initial temperature rise is quite rapid. But as the imbalance between the energy gained and lost by the pot is lessened, the rate of warming slows. Given enough time, a new, higher equilibrium temperature is eventually reached where the rate of energy gain equals the rate of energy loss. This is true of a car sitting in the sun, too. Its temperature will rise rapidly when it is exposed to sunlight, but will eventually stop rising as the rate at which solar energy is absorbed is matched by the rate at which the hot car loses energy to its cooler surroundings.

It is important to understand why the temperature response curves in Fig. 10 look the way they do. As the temperature

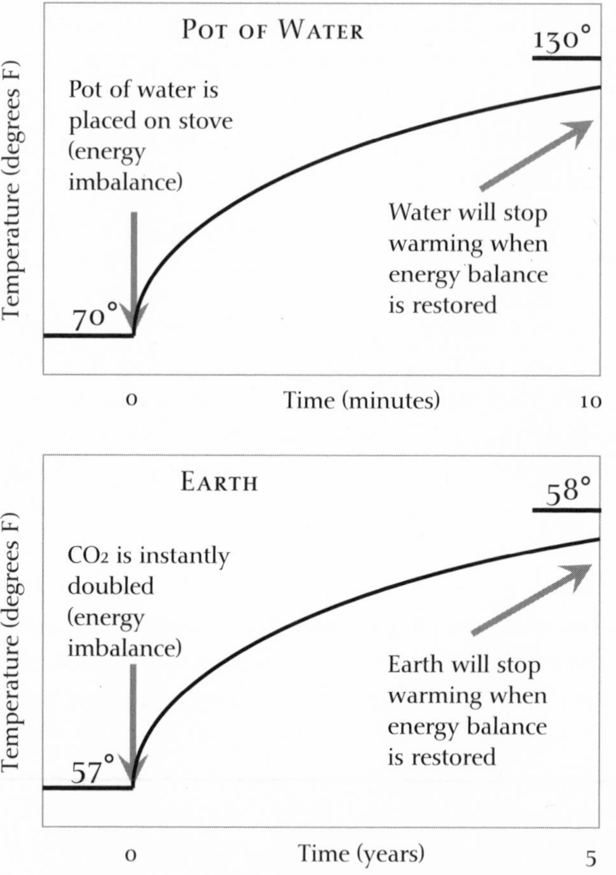

Fig. 10. Whether it is a pot of water on the stove, or the Earth, warming as a result of an energy imbalance occurs more rapidly at first, then more slowly as a new state of energy balance is approached.

responds to an energy imbalance, a portion of the imbalance has then been relieved by the temperature change. This means that the remaining energy imbalance is smaller, which in turn causes a smaller temperature response, which decreases the imbalance even more. This process is continuous, with progressively smaller imbalances causing progressively smaller temperature changes. The final result is a temperature change with time that looks like that in Fig. 10.

As can be seen in the lower panel of Fig. 10, the Earth's temperature response to an energy imbalance is qualitatively the same as that seen with the pot of water. But because there is so much "stuff" on Earth to heat up, it takes a lot longer for the temperature rise to occur. For most of us living in the Northern Hemisphere, for instance, the maximum amount of sunlight available occurs at the summer solstice, around June 21. But it takes time for that extra energy to warm the land and the atmosphere, so the peak in temperature doesn't occur until about two months later, in August.

In the context of global climate change, the role of the oceans is more important than the land because they can store so much more heat, and therefore take a lot longer to warm or cool. While the heat-carrying capacity of the atmosphere is equivalent to only about 2 meters (about 6 feet) of water, temperature changes in the ocean are spread over tens of meters on time scales of days, and to hundreds of meters over many years. Thus it can take years, if not decades, for the world's oceans to warm or cool in response to an energy imbalance.

There is still considerable uncertainty over how long it takes the Earth to fully respond to a forcing (energy imbalance) because we are not sure how deep into the ocean the temperature change in response to an energy imbalance can reach. In fact, it is safe to say that the Earth never does fully respond to a forcing because the energy imbalance is always changing anyway. The temperature is always trying to play catch-up, with a maximum in forcing preceding the maximum in temperature, so that true energy balance is never reached – except maybe briefly by accident. It's like turning the heat up and down on the stove every few minutes: the pot of water will never reach energy balance and a steady temperature.

Because some of the radiative imbalance from humanity's greenhouse gas emissions has already been relieved by the global warming we've experienced over the last century or more, the imbalance of 1.6 watts per square meter that I mentioned above is not what is believed to exist today. James Hansen has calculated the remaining energy imbalance in the global climate system to be about 0.8 watts per square meter.[5] This remaining radiative forcing represents what is commonly called global warming "still in the pipeline," which means warming that has not yet occurred but will happen in the coming years even if humanity immediately stops all greenhouse gas emissions today.

Of course, we are still adding greenhouse gases to the atmosphere. So the global energy imbalance will keep increasing for many years to come, and the resulting warming will take even longer to be fully expressed.

But Hansen's calculation of the warming "still in the pipeline" assumes something very important. It assumes that there are no significant natural sources of warming or cooling, such as from a natural change in global cloud cover or atmospheric water vapor. Since our satellite instruments are not accurate enough to confidently measure a global energy imbalance smaller than about 2 or 3 watts per square meter, we really do not know what the energy imbalance of the Earth is. That Hansen considered his calculation of the remaining radiative imbalance of the Earth to be "smoking gun" evidence shows how quickly scientists who work in the realm of theory forget the assumptions they made along the away–or didn't even know they were making.

If something like natural cloud variations is the cause of most global warming, then Hansen has to be wrong on the holy grail of all climate unknowns: *feedbacks*. If Hansen is right and feedbacks are positive, then the warming we've experienced in the last fifty years probably can be explained by our greenhouse gas emissions alone. But if feedbacks are negative, then our greenhouse gas emissions are not enough to explain the warming.

There must be some natural mechanism at work causing most of the warming.

Which brings us to the last and most important question: *What are the feedbacks in the climate system?* The answer to that question will determine how much anthropogenic warming we can expect in the future.

Chapter 4 · Feedback: How Much Warming Results from the Forcing

While forcing (an energy imbalance) determines whether a temperature change will occur, feedback determines how big that temperature change will be. It is feedback that ultimately determines whether man-made global warming is catastrophic, or merely lost in the noise of natural climate variability. There are familiar examples of feedback all around us every day.

WE HAVE DISCUSSED one main component of climate change: the extra carbon dioxide we put into the atmosphere causes an energy imbalance (*forcing*), which then results in a warming tendency. I currently agree with the IPCC on this part of the global warming problem. It is the second component, *feedback*, that is the source of most scientific debate over global warming, and it is the most important part.

Feedback refers to how clouds and other elements of the Earth's climate system change in response to a temperature change, thereby either magnifying or reducing that temperature change. In the context of global warming, positive feedbacks will magnify a warming tendency from the extra CO_2, while negative feedbacks will reduce it. Note that feedbacks are always referenced to temperature.

If the sum of all feedbacks in the climate system is strongly positive, then catastrophic global warming can result. It might be expected to be accompanied by other dramatic changes in climate:

droughts, floods, hurricanes. I've found that the belief in strongly positive feedbacks is held by only a minority of climate scientists, although most researchers would probably agree that catastrophic warming cannot be ruled out. But those with the most extreme views on feedbacks also end up being the ones who are most often quoted by the news media. Remember, bad news is good news.

If the sum of all feedbacks is negative, on the other hand, manmade global warming will not be a serious concern. Remember, the direct warming effect of the extra carbon dioxide is agreed by everyone to be small. It's the indirect warming that has many experts concerned. Negative feedbacks would end up reducing the small amount of direct warming from the extra carbon dioxide still further, maybe to the point of being unmeasurable. Scientists who are not predicting planetary doom are not going to make for very interesting news stories, though, so you will not hear about it when our scientific papers are published.

Now, for the engineers, physicists, and science-savvy among you, there might be some confusion at this point regarding the term "feedback" in the context of climate change. In all disciplines other than climate, "positive feedback" is understood to cause a system to become unstable. The most familiar example is the screech you hear from a public address sound-amplification system when there is positive feedback of the sound coming out of the speakers, feeding back into the microphone to be amplified still more. A positive feedback loop is set up, and the system becomes unstable. But in the realm of climate research, even the IPCC's climate models remain stable and do not cause runaway global warming, although they all exhibit positive feedbacks.

The inconsistency is due to the fact that in climate work we do not consider the extra infrared energy that results directly from a temperature increase to be part of feedback, even though it acts like negative feedback. It is understood that the Earth will always give off more infrared energy to space as it warms, which then keeps the climate system stable.

But positive feedbacks will counteract a portion of that extra infrared energy being given off, causing the system to warm even more in order for energy balance to be restored. Negative feedbacks, in contrast, will enhance the loss of extra infrared energy, which means the Earth will not have to warm as much to restore energy balance.

This is why, in the realm of climate change, positive feedbacks do not cause an unstable system. It's just a matter of semantics and creative accounting of the energy flows.

SIMPLE EXAMPLES OF FEEDBACKS

The concept of feedback is something you already understand from your everyday experience. Let's take the example of a car sitting in the morning sun (Fig. 11). Sunlight shining into the car causes an energy imbalance (a forcing), and so the temperature inside the car begins to rise. If you were inside the car, experiencing the rising temperature, you would probably open the window. Voilà! You have just turned yourself into a negative feedback mechanism, sensing the temperature increase and then changing

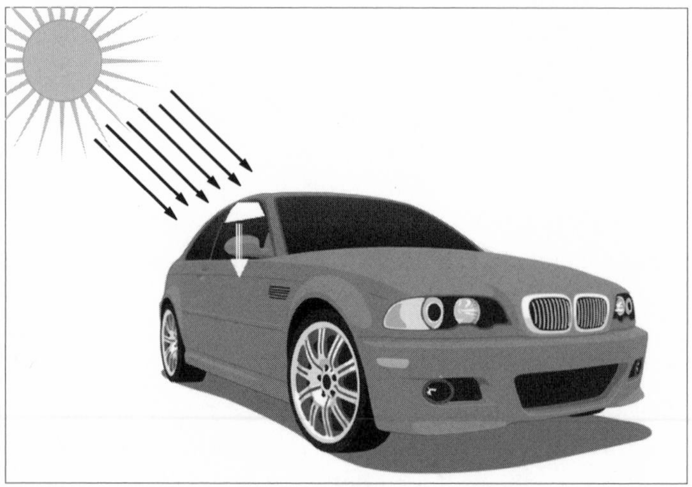

Fig. 11. When you open a car window as the interior warms in the sun, you have created negative feedback, which then acts to reduce the magnitude of the temperature increase.

the system in such a way as to reduce the magnitude of the increase.

This is exactly analogous to how negative feedbacks would operate in the climate system. As we will see in the next chapter, our latest and best satellite observations suggest that this is the way the Earth's climate system works, with negative feedbacks acting to reduce any temperature increase forced upon it.

In contrast, all IPCC climate models behave in just the opposite way: with positive feedbacks. When a temperature change is forced upon the models, they react by amplifying it. This would be like sitting in the car with the window open at first, and then as you feel the inside get warmer, you gradually close the window to make the temperature climb even faster.

If feedbacks in the climate system are negative, we have little to worry about in terms of future manmade warming. But it also means that the extra carbon dioxide in the atmosphere is not sufficient to explain past warming, either. There would have to be some natural warming mechanism at work. I will discuss new evidence for that possibility in greater detail in Chapter 6. A related question is why, if nature displays negative feedback, the climate models instead exhibit positive feedback. After all, climate models are constructed with nature as a guide, so why would they behave unrealistically? I will explain the reason for this dichotomy in the next chapter.

There are other examples of negative feedback mechanisms all around us. Remember that your body has a thermostatic mechanism for warming it when it gets too cold (shivering), as well as for cooling when it gets too hot (sweating). The thermostat controlling your home heating or cooling system is also part of a negative feedback mechanism. When the thermostat senses that the temperature has become too high or too low, it turns on the air conditioner or the heater to bring the temperature back closer to a preset value.

What would constitute a negative feedback mechanism in the case of the pot of water heating up on a stove? Some pots have a relief valve in the lid to release extra heat once the water begins to boil (see Fig. 12). This relief valve would be considered a negative feedback mechanism because it increases the rate of heat loss while warming is occurring. Let's assume that the relief valve springs into action even before the water boils, opening more and more as the water gets warmer. In this case, energy balance will be reached at a lower temperature than if the relief valve were not there. This situation, which is analogous to a car window being opened to reduce the temperature increase inside, is shown in Fig. 12. A positive feedback, in contrast, would require the relief valve to do just the opposite: to close up as the pot grows warmer, trapping more heat and increasing the temperature even further.

Feedback determines the amount of warming that results from an energy imbalance.

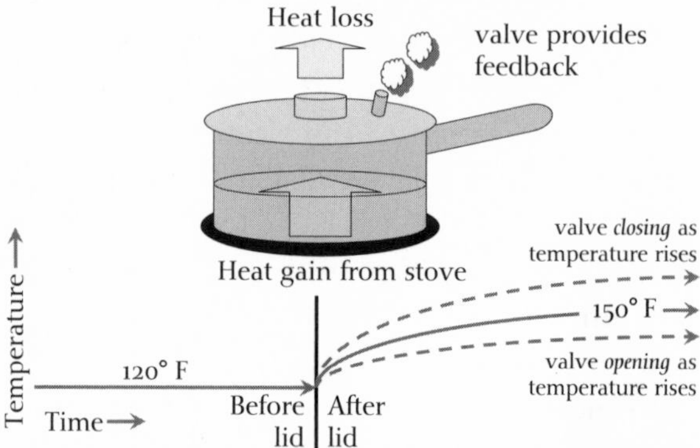

Fig. 12. Negative feedback in the climate system would be like a relief valve in the lid covering a warming pot of water, opening wider as the temperature rises. Positive feedback would correspond to the relief valve closing as the temperature rises, thus amplifying the warming.

ESTIMATING FEEDBACKS

Returning to the example of the car, we will next examine how researchers have tried to estimate feedbacks in nature from satellite measurements of the climate system. For reasons that will become apparent later, we will assume that the car window is half-way open to begin with. The sun starts to shine in, which causes a forcing that makes the inside of the car grow warmer and warmer. If you slowly open the window more as the temperature in the car rises, you have created a negative feedback mechanism.

This situation can be represented in the form of a graph (Fig. 13), with temperature on the horizontal axis and heat lost through the window on the vertical axis. I am introducing the graph because it is similar to the way researchers plot satellite data.

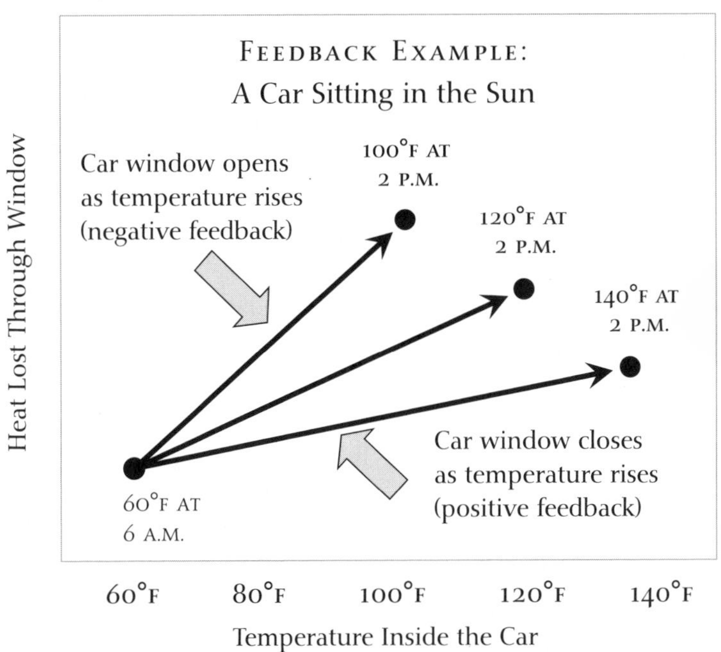

Fig. 13. Feedback can be estimated by plotting changes in temperature versus heat loss on a graph, and drawing lines to connect the dots. Strongly sloping lines suggest negative feedback, while slightly sloping lines are characteristic of positive feedback.

The data point in the lower left corner represents the situation before the sun starts shining in, with a temperature inside the car of 60 deg. F and the window halfway open. If we gradually open the window further as the temperature rises in the car, we will eventually reach the point where a maximum amount of heat is being lost through the open window, and the temperature increase is kept to a minimum, let's say 100 deg. F. This transition is represented by the most steeply sloping line in Fig. 13, and it corresponds conceptually to the negative feedback case.

If we instead gradually close the window as the temperature rises, the inside of the car will reach a much higher temperature, say, 140 deg. F, and a smaller amount of heat will be lost through the window. This is the positive feedback case, and is represented by the line in Fig. 13 with the least slope.

The line between the positive and negative feedback cases represents the zero feedback case. The window remains halfway open no matter what the temperature inside the car is. In this case there is still more heat lost through the window with warming, but the window itself does not "feed back" on the temperature increase by either opening or closing. In the climate system, the zero feedback case corresponds to the direct warming effect of extra carbon dioxide in the atmosphere. As mentioned previously, this direct surface warming from a doubling of atmospheric CO_2 would be about 1 deg. C by the end of the twenty-first century.

In the case of the car, one could envision other ways to accomplish the feedback. For instance, putting a sun shade in the windshield to reduce the amount of sunlight entering the car would also be a negative feedback on the warming.

Since we will be using similar graphs when analyzing the satellite measurements, it is essential to understand that it is the *slopes* of the lines shown in Fig. 13 that are assumed to be a measure of feedback. Therefore, anything that affects the line slope will affect our interpretation of feedback.

Negative feedback produces the most steeply sloping line, indicating a maximum amount of heat lost for a given rise in temperature. This is how I believe the real climate system works, with a warming influence causing a maximum amount of excess heat loss to outer space. In effect, the Earth "opens a window" to outer space (or puts up a sun shield) as the temperature rises in response to a forcing, allowing more energy to escape. The result is a smaller temperature rise in response to the forcing than if there were no feedback, or positive feedback.

But the IPCC climate models, in effect, close that window with warming, which causes a greater temperature increase in response to a forcing. In fact, all of the IPCC climate models exhibit net positive feedback, like the least sloping line in Fig. 13.

Potential Positive Feedbacks in the Climate System

Al Gore's and James Hansen's apocalyptic view of global warming depends entirely on the alleged existence of strongly positive feedbacks in the climate system. In the atmosphere, these positive feedbacks could include more water vapor, which is our main greenhouse gas; less coverage by low clouds, which would let in more sunlight; and greater coverage by thin, high-altitude cirrus clouds, which would trap more infrared energy.

In all IPCC climate models the sum of these atmospheric feedbacks is positive, increasing the rate of temperature change in response to a forcing, and thereby leading to forecasts of dangerous levels of global warming. James Hansen has expressed great concern over the melting of the Greenland ice sheet and a resulting rise in sea level by many feet, something that could occur only (if at all) with strongly positive feedbacks.

Potentially the largest and yet the most uncertain feedback is that due to clouds. I cannot overstate the importance of the uncertainty over cloud feedbacks. At least theoretically, clouds could either save us from global warming, or cook us. The new satellite evidence I will present directly addresses the cloud feedback issue.

There are non-atmospheric feedbacks, too. Probably the most familiar surface feedback is related to the melting of snow, glaciers, and sea ice. As snow and ice melt, their bright reflective surfaces are replaced by dark water or land that was underneath the ice or snow. This would be positive feedback because dark surfaces absorb more sunlight than bright surfaces as the snow or ice melts. Fortunately, these frozen surfaces cover a relatively small fraction of the Earth, and exist in regions or times of year that receive relatively little sunlight anyway.

While the scientific consensus is that these are positive feedbacks, it is also possible that in a warmer world there would be more precipitation on the Greenland and Antarctic ice sheets.[1] Since temperatures over these ice sheets are almost always too cold for anything but snow to fall, the ice sheets could actually grow rather than shrink with warming. So, there are complex and competing processes that do not permit an easy interpretation of whether this feedback is positive or negative when it comes to the huge landlocked ice covering Greenland and Antarctica.

If you are wondering about surging glaciers in Greenland dumping more ice into the ocean and huge ice shelves in Antarctica breaking off, these are things that have always happened and always will happen. As long as snow continues to fall on glaciers and ice sheets, they will continue to flow into the ocean. Antarctica is ringed by ice shelves sticking out over the ocean, and as long as snow continues to fall on the continent, ice shelves will periodically break off and float away. While it is easy for researchers to speculate, there is no convincing evidence that recent changes in glaciers and ice shelves are in any way due to the activities of humans. For instance, newly published evidence suggests that the recent surging of glaciers and increased outflow of melt water in Greenland that has been a focus of so much concern has stopped, probably the result of a natural fluctuation in glacier behavior.[2]

I believe that snow and ice feedbacks will end up being a largely irrelevant issue in global warming anyway. Feedbacks from the melting of the Greenland and Antarctic ice sheets would take a very long time to materialize, they are regional, and they would occur at relatively high latitudes where there is little sunlight. In contrast, atmospheric feedbacks are rapid, and they occur everywhere on Earth.

This means that if atmospheric feedbacks are negative, then any positive feedbacks from melting ice and snow won't have a chance to occur in the first place. Feedbacks are, by definition, responses to a temperature change, and if atmospheric feedbacks do not allow warming, then there will be no feedback from melting snow and ice. Partly for this reason, I will be focusing on atmospheric rather than surface feedbacks.

One of the undesirable characteristics of multiple positive feedbacks is that they amplify each other. Multiple positive feedbacks interacting is like taking a number and doubling it, then doubling it again, then again, and so on. You approach very large numbers very quickly. The same multiplying effect happens in the case of multiple negative feedbacks; but instead of being like doubling a number, it's like halving a number, then halving it again, and then again. You slowly approach zero, but never quite reach it. As a result, there can be no climate catastrophe with net strongly negative feedbacks–only with net strongly positive feedbacks.

So this is the crux of the most important question that still exists in global warming research today: *What are the real feedbacks in nature?* The answer to that question will determine whether mankind now controls climate, or whether the effect of humanity is just a small bump on the long, hilly road of natural climate variability.

Next we will examine how feedbacks have been estimated from satellite measurements of natural climate variability. In the process, we will discover a fundamental mistake that researchers have made.

Chapter 5 · How Mother Nature Fooled the World's Top Climate Scientists

A mix-up between cause and effect in observations of cloud behavior from satellites has led to the false illusion that our climate system is dominated by positive feedback. This, in turn, has led to the development of highly sensitive climate models that predict large amounts of global warming. But when the separate influences of forcing and feedback (cause and effect) are isolated, recent satellite data reveal the climate system to be dominated by negative, not positive, feedback.

IN THIS CHAPTER and the next, I take my case to the people. I have virtually given up hope that the climate research community will objectively address the subject of natural sources of climate change anytime soon. The Keepers of All Climate Knowledge have erected a nearly impenetrable barrier to any new science that does not support the current paradigm of anthropogenic global warming, as defined and guided by those controlling the IPCC process. Published research that should be causing the climate modeling community to sit up and listen is instead being ignored. Groupthink has taken over.

If the climate system is insensitive, we are wasting effort in trying to reduce our emissions of greenhouse gases. Yet the U.S. government is poised to make some highly consequential policy decisions with that purpose in mind. The House of Representatives has already passed cap-and-trade legislation to regulate carbon dioxide emissions by businesses. The EPA may regulate CO_2 production now that the Supreme Court has told them to consider it a "pollutant." At least one of these supposed "remedies" for global warming may already have been implemented by the time you read this book.

Such policies would increase the price of virtually all goods and services, because everything we do requires energy. If you penalize energy use, you destroy wealth, and when wealth is destroyed, the poor are the first to suffer. Again, I do not really care where our energy comes from, but I do care that it be as inexpensive as possible, making all the goods and services that require energy more affordable as well. This enables people to lift themselves out of poverty and attain the health and standard of living that many of us now take for granted. Some radical environmentalists even say we should prevent the poor of the world from building wealth, thereby saving the Earth by keeping the poor from making the same mistakes we have supposedly made. But I cannot in good conscience support any position that keeps people from reducing mortality and making their daily lives easier.

Legislation and regulation resulting from fears of anthropogenic climate change are supposed to be based on the best available science. But when scientists like me publish research that goes against the consensus, it is simply ignored by both the research community and the news media. That is what happened to our two most recent published papers that cast doubt on the IPCC view of manmade global warming.[1] We are not the only researchers to have had this kind of experience with published papers that did not align with the dominant view. This situation allows climate alarmists like Mr. Gore to assert that there are no real climate scientists who dispute the scientific consensus. The public is left with a biased impression of the state of the science, because they have never heard about important work that has indeed been published. And it would take only one research study to cause the global warming house of cards to collapse.

* * *

THE CLIMATE RESEARCH ENVIRONMENT

I do not believe that there is any widespread conspiracy among the scientists who are supporting the IPCC effort – just misguided good intentions combined with a lack of due diligence in scientific research. I've heard statements from scientists like, "Getting off fossil fuels is the right thing to do anyway." While this reveals a bias, it hardly constitutes a conspiracy.

But I do believe that there is a conspiracy among some politicians and some of the IPCC leaders to get international agreements to regulate greenhouse gas emissions *no matter what the science says.* Whether their motivations are financial or political, or they are looking for meaning in their lives, these folks seem to be hoping that humanity is in grave danger from manmade global warming. One reason why the scientific community goes along with this is that they know there is at least the possibility that dire predictions of global warming and climate change *might* come true. Scientific investigation always involves uncertainty, and this is especially true for climate research.

A variety of dynamics have resulted in an overall lack of scientific objectivity in the climate research community. Everyone in this business has biases on the subject of global warming. The bias starts from the very beginning with congressional funding of scientific research. For those who think government-funded research is impartial, I can tell you from firsthand experience that it is not.

In order to convince Congress to fund research into a problem, you must first convince them that a problem exists. This automatically makes manmade global warming a particularly lucrative field for funding – as long as the threat of manmade global warming continues. There are managers at NASA, the National Oceanic and Atmospheric Administration, the National Science Foundation, and the Department of Energy whose careers now depend on a continuous flow of research dollars through them to the science community. As research programs are built and careers established, an entrenched scientific constituency develops. Scientists have to support their families, and the older we get, the more difficult it is to change fields of research. Note that I include myself in this group, since my research is funded entirely by the U.S. government.

Then there are the huge political implications of mankind being in control of the climate system. Political power derives mostly from control over the public purse, so the global warming issue is perfect for those whose careers depend on deciding how our tax dollars should be spent. And since everything we do requires energy, whoever controls the world's energy supply controls the world.

There is also the arranged marriage between politics and science, something that would not have come about naturally. The IPCC was formed over twenty years ago largely for political reasons: to build the scientific case that mankind causes global warming, and thus the policy case for regulating carbon dioxide emissions. Because almost all options for tackling global warming involve more governmental control over society, a political bias ends up coloring the IPCC leadership's message in a way that minimizes scientific uncertainties and maximizes public alarm.

While many people in government are true public servants, dedicated to improving others' lives and not just their own, plenty of them are also bent on increasing their own power and wealth, and they see the expansion of government control over our lives as a means to that end. Controlling energy use would be an enormous expansion of government power, so these people pursue it not just for scientific reasons but also for self-advancement.

Finally, there are the less tangible motivations that can result in research bias. For instance, what scientist wouldn't want to work in a field where he could help Save the Earth?

These are some of the factors that have led climate scientists to accept a manmade explanation for climate change too readily and too uncritically. In short, the issue of global warming involves scientists and politicians who all have a vested interest in the consensus being correct. This has led to a research environment that is incentivized to avoid any lines of research that might cast doubt on the idea of humanity as the primary cause of global warming.

Am I saying that the funding of global warming research is a bad idea? No. I think we have much more to fear from natural climate change than from anthropogenic change, and that is just as worthy of study. But global warming research unavoidably involves the creation of bureaucratic and scientific infrastructures and constituencies whose number-one job is to keep research funds flowing. There is a built-in bias to keep the threat of manmade global warming alive for as long as possible. The result is not so much a conspiracy as it is a mindset that prevents scientists from being as objective and thorough as they might otherwise be.

FEEDBACKS ARE THE KEY

It would be difficult to exaggerate the importance of measuring whether the Earth's atmosphere "opens or closes the window" in response to an energy imbalance like that from increasing atmospheric carbon dioxide concentrations. It is the holy grail of climate research, for it will determine how serious a problem manmade global warming will be.

Feedback is the big-picture, bottom-line, end-of-the-day issue that trumps all others in the global warming debate. If we can actually measure the feedbacks operating in the real climate system, we can then easily determine how much manmade global warming will occur in response to our greenhouse gas emissions.

There are actually quite a few ways in which feedbacks have been estimated from observational data, both during the modern instrumental period over the last century or so, and from proxy estimates over thousands to millions of years.[2] But there is a fundamental problem common to all: *our estimates of past temperature change are better than our estimates of the forcings that caused them.* And measuring a temperature change without knowing what forced it is the perfect recipe for mistakenly diagnosing positive feedback.

You see, a temperature change can be caused either by a weak forcing that is being amplified by positive feedback, or by a strong forcing that is being reduced by negative feedback. Therefore, as one approaches a zero forcing, you need to have very strong positive feedback to cause an observed temperature change. Because of this issue, most methods of diagnosing feedbacks are prone to giving the illusion of a sensitive climate system. This is because we see evidence of a temperature change, but often do not know what caused it. This is just one step away from assuming that it must have been caused by some tiny forcing being amplified by positive feedback. And just like some ancient tribe of people who made sacrifices to the gods of nature to ward off severe weather, our lack of understanding of the natural forcings that caused temperature changes of the past leads us to blame our sinful use of fossil fuels instead.

For example, I discussed the Vostok ice core record of the ice ages in Chapter 2. James Hansen has claimed that the very weak forcing from the Milankovitch cycles in the Earth's tilt and orbit around the sun caused the ice ages and warm interglacial periods, in which case the climate system would have to be very sensitive, with positive feedbacks amplifying that weak forcing. But we also saw that the timing of the Milankovitch cycles relative to the ice ages was no closer to the major temperature changes than what might be expected by chance. Therefore, it is reasonable to suspect that the ice ages and the interglacial periods of warmth were caused by some as yet undiscovered forcing mechanism. Yet if one assumes that the ice age cycles were caused by the weak Milankovitch cycles, then one must also *assume* strongly positive feedbacks. I believe this is what Hansen has done: assumed an incorrect, weak forcing, and therefore he also had to assume a strongly positive feedback.

So we see that an observed temperature change must be accompanied by a good estimate of what caused it in order to estimate feedbacks. If you don't know what forced the temperature change, you will likely diagnose positive feedback.

I believe that our greatest hope for determining what feedbacks are operating in today's climate system is by actually measuring today's climate system. If we cannot figure feedbacks out from actual measurements of how the climate system operates today, how can we ever hope to rely on past events like ice ages when we have no direct measurements of those events to analyze? Given the importance of the search for this holy grail of climate research, you would think we'd have flocks of researchers examining reams of satellite data to ferret out what nature is trying to tell us regarding forcing and feedback. But as far as I know, I am the only one who is currently examining this issue in great depth. I say "in great depth" because, as we will see, other researchers have been fooled when they were not careful enough about what they were seeing in the satellite data.

Because of too little due diligence, the determination of feedbacks has remained stubbornly difficult in previous research efforts. Various investigators have used different satellites, analyzed different periods of time, and arrived at different answers. The reasons for these disparities have remained unclear, but the explanations that have been offered are bad satellite data or feedbacks changing with time. The most critical review of feedbacks ever published also concluded that we need new and better methods to measure feedbacks in the climate system reliably.[3]

The lingering ambiguity over observed feedbacks is the biggest reason for the large range of global warming projections that continue to come out of the IPCC. For instance, the IPCC in 2007 claimed over 90 percent confidence that global warming in response to a doubling of carbon dioxide will not be less than 1.5 deg. C (2.7 deg. F), and that warming greater than 4.5 deg. C (8.1 deg. F) could not be ruled out. This is a very wide range of warming estimates, corresponding to anywhere from modest to potentially catastrophic climate change.

Imagine if, instead of a wide variety of feedback estimates, previous researchers always obtained the same feedback from our satellite measurements. Then climate modelers would have little choice but to make sure their models produced about the same result. In that case, their climate models would all predict roughly the same amount of warming.

But this has not been the case. It has even been suggested that the task of measuring feedbacks is hopeless, and so we should give up.[4] And it is this persistent uncertainty that keeps the fear of catastrophic global warming alive. As long as scientists can claim there is a great uncertainty about feedbacks, then there is always the possibility that the climate system is extremely sensitive and that we have already pushed the Earth past the point of no return. Since feedbacks are so uncertain, it is argued, we have to accept the possibility that climate sensitivity is very high, dominated by strongly positive feedbacks, which would mean we are in for really big trouble. This uncertainty then provides the ammunition that politicians and environmentalists need for proposed policies to limit carbon dioxide emissions. Better safe than sorry, it is argued.

Cloud Illusions

The mistake I claim researchers have made in estimating feedbacks involves the role of clouds in the climate system. The evidence I will present deals with the relationship between clouds and temperature that we observe with natural, year-to-year climate variations. Returning to the analogy of a car warming in the sun: there has been a mix-up by researchers who were trying to determine whether clouds "open the window" or "close the window" during warming.

It might sound a little hokey, but climate researchers' misunderstandings over the role of clouds in the climate system remind me of a verse from Joni Mitchell's song *Both Sides Now*, made famous by Judy Collins in 1968:

I've looked at clouds from both sides now,
From up and down, and still somehow
It's clouds' illusions I recall—
I really don't know clouds at all.

I will argue that scientists have been fooled by an "illusion" because they have not "looked at clouds from both sides." As a result, they "really don't know clouds at all."

The two sides of clouds' role in global warming involve forcing and feedback—or if you prefer, cause and effect. Forcing (cause) would be clouds causing a temperature change. Feedback (effect) would be causation flowing in the opposite direction, with temperature causing a cloud change. This effect then feeds back upon the original temperature change, making it larger or smaller.

The best way to describe the "illusion" that I believe affects climate researchers when it comes to cloud behavior is to relate how I got involved in this line of research in the first place. I had always heard that climate models produce large amounts of global warming on account of positive feedbacks in those models. When I asked for the evidence that positive feedbacks really exist in nature, I would be told that our satellite observations showed that there was, on average, less cloud cover over the Earth in unusually warm years. Therefore (the argument went) the warming caused less cloud cover, which allowed more sunlight in, which enhanced the warming. This observation was given as an example of positive feedback in nature.

But something bothered me about this explanation. How did the researchers know that the warmer temperatures caused a decrease in cloud cover, rather than the decrease in cloud cover causing the warmer temperatures?

Well, it turns out they didn't know.

Along with the computational physicist who works with me, Danny Braswell, I decided to investigate this cause-versus-effect issue with a simple climate model. We found that clouds causing a temperature change could give the illusion of positive feedback even when we specified negative feedback in the climate model.[5] In other words, if researchers are not careful about distinguishing cause and effect when observing cloud and temperature variations, they can be fooled into believing that the climate system is more sensitive than it really is. Notice that all I did was apply intuition along with my understanding of weather to ask a basic question about cause versus effect: Do clouds cause temperature to change, or does temperature cause clouds to change? As far as

I have been able to determine, no one else has ever asked this question.

And this leads me to another reason why I wrote this book. The fundamental problem of causation was one that I found the public understood better than the scientists did.

We submitted the results of our study for publication in what is arguably the world's leading scientific journal for climate research, the American Meteorological Society's *Journal of Climate*. I did not have high hopes for getting the paper accepted, though, because of its potential implications regarding the seriousness of manmade global warming. To my great surprise, two leading climate experts chosen by the journal's editor to be peer reviewers agreed that we had raised a legitimate issue. In fact, each reviewer decided to build his own simple climate model to demonstrate the effect for himself. Both offered constructive advice on how to improve our model in order to demonstrate the effect more clearly. One even said it was important that the climate modeling community be made aware of the issue. We modified the paper according to their advice, and it was published in November 2008.

Our university put out a press release on the paper—and the mainstream news media totally ignored it.

As far as I can tell, the results of that published work have been largely ignored by the scientific community too. Chances are, even if they did read the paper they would not recognize its potential significance. This is because it is almost impossible to get away with saying anything like "this could throw all of our global warming predictions out the window" in a scientific publication. There will always be at least one peer reviewer of your paper who has so bought into the theory of anthropogenic global warming that he will not permit you to publish anything that directly calls the prevailing orthodoxy into question. But for some reason it is always permissible to say something like "this means global warming could be worse than we thought." Apparently, that shows you care about the Earth.

The evidence I will present is relatively easy to understand, with examples from everyday life, even if you weren't very good at science or math in school. If you are not interested in the details or you have a violent aversion to graphs, then go directly to the "bottom line" summary toward the end of the chapter. But even if you are an experienced climate researcher, there will be enough "meat" here to challenge what you thought you knew about the role of feedbacks in global warming.

Satellite Estimation of Feedbacks: The Traditional Way

First let's review what we are looking for in order to measure feedbacks. In the case of the car warming in the sun in Figs. 11 and 13, we would need to measure how much extra heat is lost out the window at different interior temperatures. Feedback involves opening or closing the window with warming, which will make the heat loss out the window increase either more rapidly with warming (negative feedback), or less rapidly with warming (positive feedback). In the case of the Earth, we need an estimate of how much extra radiant energy is lost to outer space during warming. That extra loss may take the form of either more sunlight reflected by clouds, or more infrared radiation emitted from water vapor or cloud changes.

In order to estimate atmospheric feedbacks in the real climate system, it is best if our satellite measurements can include as much of the Earth as possible. This is because weather systems scattered all around the world are always causing huge positive and negative energy imbalances on a regional basis. Low-pressure areas have lots of clouds and precipitation and warmer air. High-pressure systems typically have clear skies, no precipitation, and somewhat cooler air. Indeed, these large energy imbalances are part of what drives our weather. But all these regional systems are interconnected, with rising air in one region being matched by sinking air in another region. These weather systems are all part of one global, continuously overturning atmospheric circulation, and it is that global circulation over which feedbacks should be measured.

Doing so requires the unique vantage point of Earth-orbiting satellites, which allow us to measure the entire Earth so we can average out all of the regional energy imbalances in order to determine whether a global average energy imbalance exists. These satellites are in what we call near-polar sun-synchronous orbits, traveling to the north up one side of the Earth, then to the south down the other side. Each orbit takes about 100 minutes, and the Earth is slowly turning underneath the satellite's orbit. In the course of a 24-hour day, the satellite makes about fourteen orbits, and it covers virtually the whole Earth twice a day.

This kind of radiation balance data from satellites has existed only for about twenty years or so; our very best data from our most recent NASA satellites only extends back to 2000. We therefore have only a limited number of years to observe the global climate system in detail, and to try to understand what feedbacks were operating during those periods of time.

Over some multiyear period, daily satellite measurements are made of how the global average temperature varies, as well as how the radiant flows of solar and infrared energy in and out of the Earth also change. These daily global measurements are then averaged in time and plotted on graphs – as we did conceptually for the example of the car in Fig. 13. I have plotted some monthly global satellite measurements of how temperature and the Earth's energy balance have varied over time in Fig. 14. The satellite-based temperature measurements come from the Advanced Microwave Sounding Unit (AMSU) flying on the NOAA15 satellite and NASA's Aqua satellite. They represent temperature averages over the lowest several miles of the atmosphere, and are the same data that Dr. John Christy and I use for monthly monitoring of global temperatures.[6]

The satellite instruments measuring the radiative energy balance of the Earth are called CERES (an acronym for Clouds and the Earth's Radiant Energy System), now flying on NASA's Terra and Aqua satellites.[7] CERES measures variations in the two radiant energy flows involved in the Earth's energy balance: (1) how much sunlight is reflected from the Earth (which then allows us to estimate how much solar energy is absorbed by the Earth), and (2) the amount of infrared energy being emitted by the Earth to outer space. How these two radiant flows of energy vary with time tells us something about imbalances in the radiative energy budget of the Earth and how it changes with time.

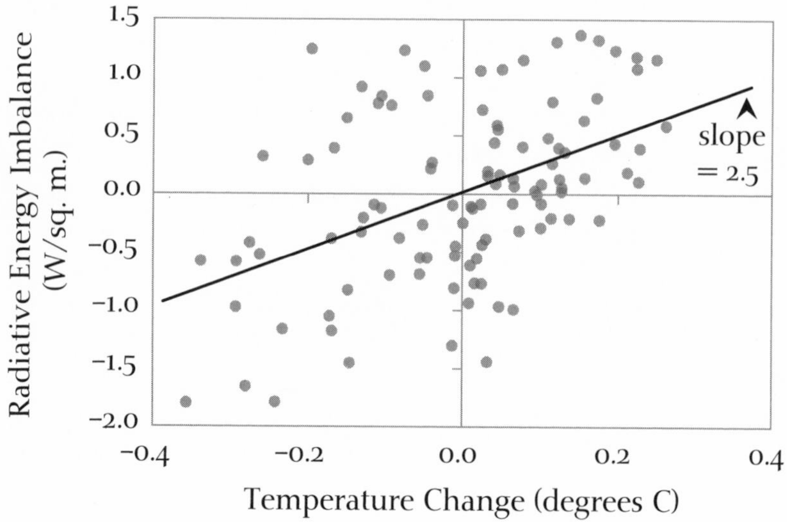

Fig. 14. Satellite-observed monthly global variations in atmospheric temperature and radiative energy imbalance averaged over the Earth from March 2000 through December 2008. The slope of the line fit to the data represents the traditional way of estimating feedback in the climate system from satellite data.

As I noted earlier, the absolute accuracy of the CERES system is not good enough to measure the tiny global energy imbalance that is supposed to result from more carbon dioxide being in the atmosphere. But the instrument is stable enough in its calibration to measure tiny *changes* in that energy balance from year to year, which are displayed in Fig. 14. This is analogous to a backyard thermometer that is sensitive to temperature changes of less than a degree, but has an absolute accuracy no better than two or three degrees.

Then, just as in the example of a car warming in the sun in Fig. 13, a straight line is statistically fit to the satellite data in Fig. 14, and the slope of that line has been assumed to be an estimate of the "total feedback parameter." It is called a *total* feedback because all the various feedbacks in the climate system contribute to it: water vapor, clouds, snow, ice, vegetation and land use, and anything else that affects the Earth's ability to absorb sunlight or lose infrared energy to outer space. The feedback parameter gives the rate at which the Earth loses radiant energy with warming, represented by the slope of the line drawn in Fig. 14, and has units of watts per square meter (of energy flow) per degree of warming. This is written notationally as $W\ m^{-2}\ K^{-1}$, where K is for Kelvin degrees of temperature change, which is the same as Celsius (C) degrees of temperature change.

In this case, that line slope, about 2.5 $W\ m^{-2}\ K^{-1}$, would correspond to weakly positive feedback. If this was the feedback operating on the warming tendency from increasing CO_2 in the atmosphere, it would result in about 1.5 deg. C (2.7 deg. F) of warming from a doubling of atmospheric CO_2. That would be at the low end of warming predicted by the IPCC.

The next illustration (Fig. 15a) shows the same monthly data as in Fig. 14, but with successive months connected by lines. Rather than just a cloud of points, we now begin to see some interesting structure emerging. The month-to-month line segments are preferentially aligned along a slope of about 6 $W\ m^{-2}\ K^{-1}$. This behavior is definitely not random, which would produce line segments pointing in random directions. Also, the slope of 6 is much steeper than the slope of 2.5 seen in Fig. 14. Might this behavior tell us something about feedbacks in the climate system?

Furthermore, if we do some smoothing of the data in time by plotting running three-month averages every month (Fig. 15b), we find some looping behavior during a brief global cooling episode that occurred during 2007–2008. A similar looping

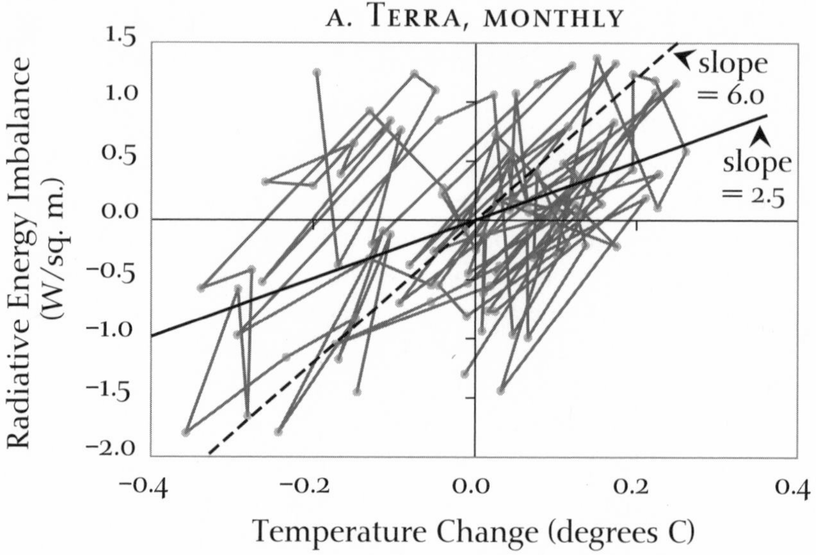

Fig. 15. (a) As in Fig. 14, but with successive months connected by lines; (b) three-month averages plotted every month and connected by lines; (c) ERBS satellite data from the early 1990s.

feature is seen after the 1991 eruption of Mt. Pinatubo in older Earth Radiation Budget Satellite (ERBS) data, shown in Fig. 15c. Because the orbit of the ERBS satellite was not sun-synchronous but drifted through 24 clock hours over a period of 72 days, 72-day averages must be used in the ERBS data analysis, rather than monthly averages.

Note also that a line statistically fit to the ERBS data has a near-zero slope, which is very different from the 2.5 slope exhibited by the more recent CERES satellite data. This is an example of why climate researchers have been so confused about feedbacks. Do we believe one satellite, the other satellite, both satellites . . . or might there be something fundamentally wrong with the way the data are being interpreted?

Clearly, the linear and looping structures revealed in Fig. 15 show that not enough thought has gone into previous diagnoses of the satellite data. Just as medical doctors rely on many years of

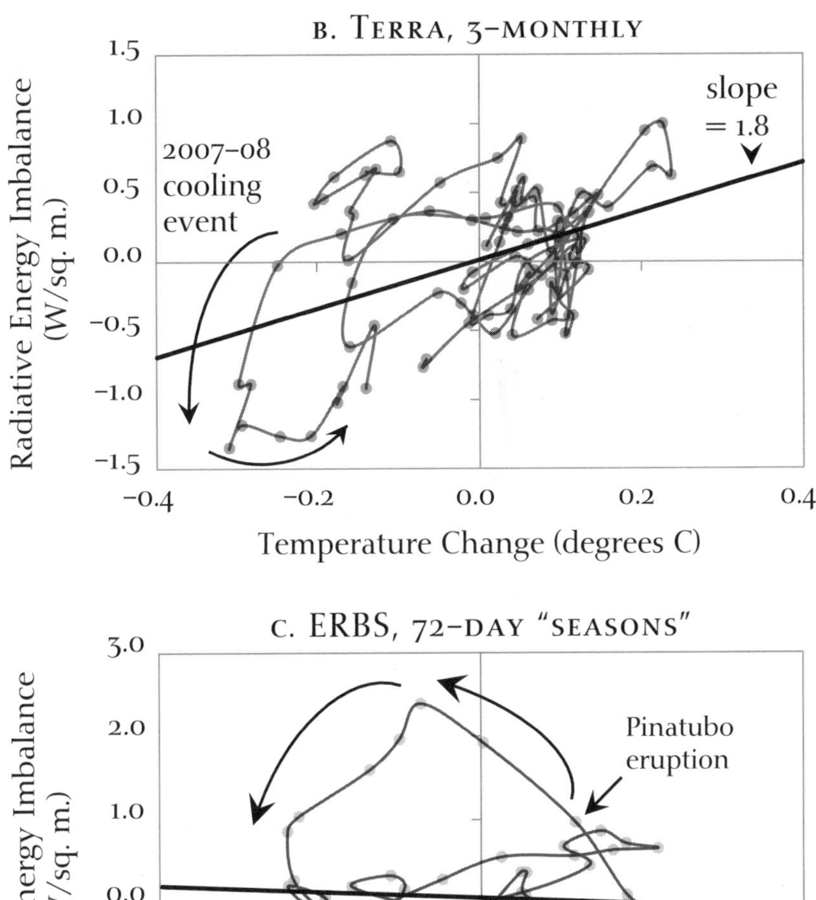

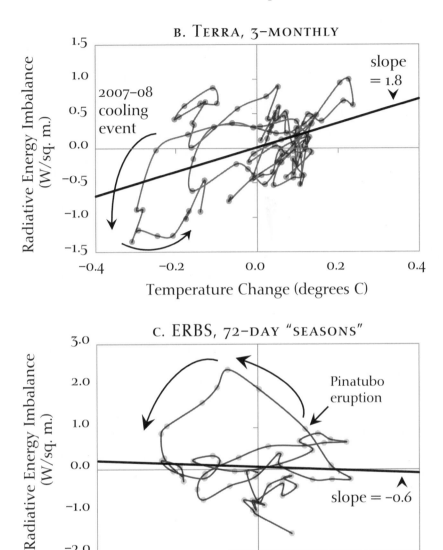

experience to interpret the subtleties of a CT or MRI scan of the human body, climate researchers must start probing deeper if we are ever to decipher what Mother Nature is telling us about the sensitivity of the climate system to our greenhouse gas emissions.

CAUSE OR EFFECT?

It turns out that a fundamental mistake has been made in previous interpretations of the satellite data. The radiative balance and temperature variations seen in Figs. 14 and 15 are due not only to feedback, but also to forcing. In other words, the behavior of the data is affected not only by temperature causing a change in the energy balance, but also by energy imbalance causing a temperature change. And since the variations in energy balance are dominated by cloud activity, what this usually represents is a mix-up between cause and effect when analyzing clouds and temperature variations.

This is the issue of causation that I have been emphasizing throughout this book: the mix-up between cause and effect when we measure natural climate variations. Returning to the analogy with the car, what we really want to know is whether the window opens or closes in response to the interior getting warmer. The trouble is, from satellite measurements of the Earth we cannot measure just the heat being lost "through the window." We can only measure the *total* energy being lost and gained by the whole car, and the feedback signal we are seeking is only a portion of that measurement. So while the sum of all the energy flows that we measure from satellites indeed gives us information on how the Earth's energy balance varies over time, we do not know how much of that imbalance is due to forcing versus feedback (cause versus effect).

To demonstrate what I am saying quantitatively, we will use a very simple (but elegant) model of the global average climate system.

A SIMPLE CLIMATE MODEL

It turns out that you don't need fancy climate models or supercomputers to do some very good global warming experiments. While the Keepers of All Climate Knowledge (the IPCC leaders) have decided that climate prediction is best left to the fastest computers, the largest number of scientists, and the most complex models, there is a very simple climate model that can be run in a spreadsheet program on your home computer to simulate global warming or global cooling.

The model is not my invention, but a simplification of the global climate system that is widely used to study basic global temperature behavior. While the model might seem almost too simple, this is deceptive. If the total feedback in the climate system is known, then the simple model can be used to predict how much manmade global warming will occur in response to the radiative forcing from more CO_2 in the atmosphere just as accurately as those big expensive IPCC climate models running on supercomputers.

What the simple model cannot do is determine how the temperature might change in different geographic regions, different altitudes in the atmosphere, and different depths in the ocean. To have any hope of doing so, the model's complexity must increase tremendously, and even the world's fastest supercomputers are not fast enough for the kinds of models that the climate modelers would like to run. But if we can demonstrate with a simple model that anthropogenic *global* warming will be minor, then who really cares how the trend is expressed on a regional basis?

If you understood the example of a car's temperature responding to an energy imbalance (forcing) from sitting in the sun, and the example of rolling the car window down to reduce the warming (feedback), you already understand the basic components of the model. It can be represented with a single equation that basically says that a change in the global average temperature with time is the result of a radiant energy imbalance (forcing) combined with the resulting radiative feedback upon that temperature change.[8]

To help our physical intuition, I have drawn the model in pictorial form in Fig. 16.

The forcing that drives the temperature change in the model can be anything that causes a global energy imbalance. Significantly, just about the only kinds of forcing that the IPCC ever names are manmade pollution, volcanic aerosols, or maybe a tiny change in the sun's output.

But what we will be most interested in is something the climate modelers have ignored: the role of natural cloud fluctuations as a forcing mechanism in addition to their role in feedback. In Fig. 16, I have drawn clouds as agents of both forcing and feed-

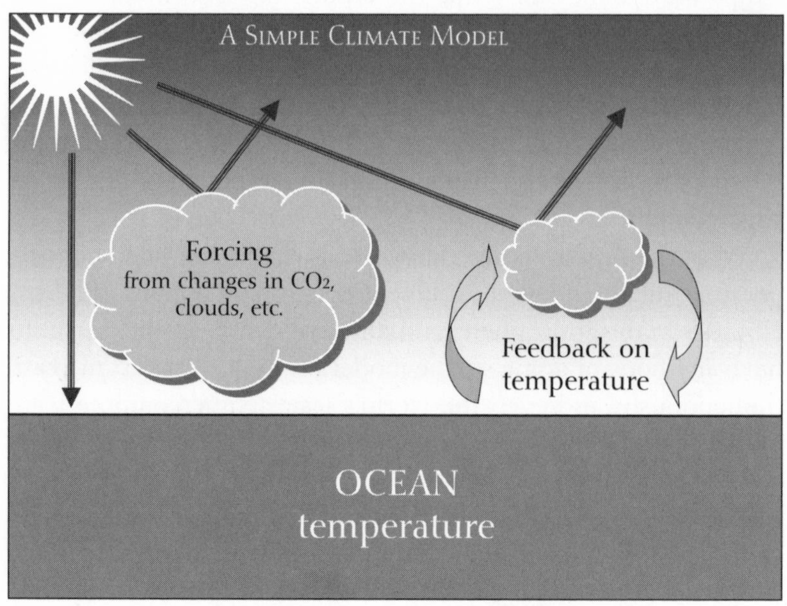

Fig. 16. The simplest climate model involves temperature deviations from average being the result of one or more forcings, combined with a resulting feedback upon that temperature change.

back, and it is the separation of these two that is the crux of the problem I am addressing. As I mentioned earlier, if we see clouds dissipate with warming, is that a sign of positive feedback, since warming causing clouds to dissipate would let in more sunlight, thus resulting in even more warming? Or was the warming caused by dissipating clouds? While the satellite can measure only the combined effect of these two processes, forcing and feedback, what we are trying to estimate is just the feedback portion.

We will now use this simple model to demonstrate the mistake that researchers have made in diagnosing feedbacks. It is easy to make mistakes of interpretation in scientific research. Measurements are the easy part of scientific investigation; figuring out what those measurements mean in terms of how nature works is the hard part.

I put equations representing the basic processes shown in Fig. 16 in an Excel spreadsheet as my model of temperature variability over time. In order to run the model we need to input estimates of: (1) the depth of ocean over which heating or cooling will be distributed; (2) an energy imbalance to force a temperature change; (3) a total feedback parameter to respond to the temperature change; and (4) a starting temperature.

This model of the global climate system assumes that forcings cause temperature deviations away from some "normal" or average state of energy balance. Feedbacks can be thought of as controlling how strongly the system tries to push the temperature back to its normal state. While one might legitimately ask whether such a simple model can realistically represent variability in the climate system, we will see that it really does reproduce behavior seen in the satellite data plotted in Fig. 15.

If I want to compute a temperature change once a month in this model, I need one line of computations per month in the spreadsheet program. Because Excel is limited to 32,000 lines of input, this means that a little over 2,500 years of climate change can be modeled with a monthly time step. Not bad for a home computer.

While the forcings that the climate modelers use change slowly in time, such as slowly increasing carbon dioxide concentrations in the atmosphere, the forcing I will use is monthly random fluctuations in energy imbalance, which are then smoothed in time. These can be thought of as random, chaotic variability in low cloud cover causing changes in the amount of sunlight being absorbed by the ocean. And just as turning the flame up or down when heating a pot of water will cause different water temperatures, different amounts of sunlight being absorbed by the ocean also causes temperature changes.

An example of the kind of temperature behavior you can get with this model is shown in the top panel of Fig. 17, which indicates the time variations of the first ten years of model radiative forcing and the model's temperature response. As the randomly varying energy imbalance evolves over time, the temperature of the climate system is continuously readjusting in response to that imbalance, while the feedback (specified as strongly negative in this run, with a value of 6 W m^{-2} K^{-1}) is always acting to reduce the temperature change and push it back toward zero.

Note that the maximum forcing always precedes the maximum temperature response, typically by about six months or so. This time lag occurs for the same reason that a pot of water takes time to heat up after you turn the stove on, as was demonstrated in Fig. 10. But whereas a few inches of water in a pot take only a few minutes to heat up, in this model experiment the water was assumed to be 50 meters (about 160 feet) deep, which takes much longer to respond fully to the forcing.

If we then plot all ten years of model output in the same manner as in Figs. 15b and 15c, shown in the bottom panel of Fig. 17, we see looping patterns like those in the satellite data. But unlike the satellite data, there is virtually no slope to the data. Despite the fact that I specified a feedback parameter of 6 in the model experiment, the slope of a line fit to the data is only 0.6.

A feedback parameter approaching zero corresponds to positive feedback so strong that a borderline unstable climate system would result. In such a climate system, any forcing at all would cause large amounts of global warming or cooling, with the global average temperature meandering around in what is called a "random walk." This is even more sensitive than any of the IPCC climate models.

How could this be? In the model run I specified a strongly negative feedback, which is represented by the dashed line I've drawn in Fig. 17, but there is no indication that the data points are clustering along a sloping line. When I statistically fit a line to the data, I get almost no slope at all – a nearly horizontal line.

This discrepancy between the specified feedback and the

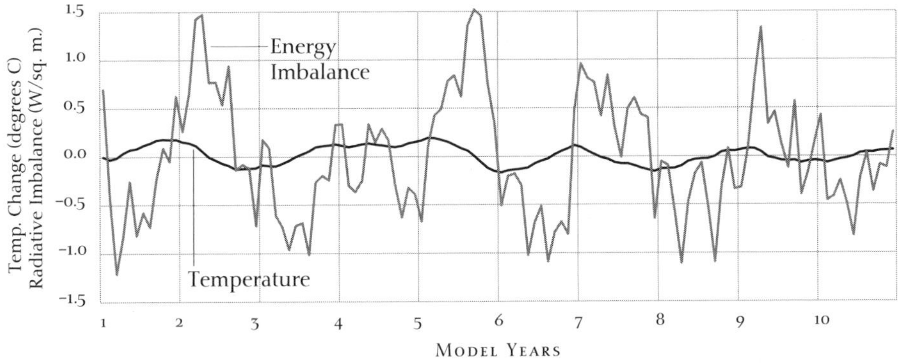

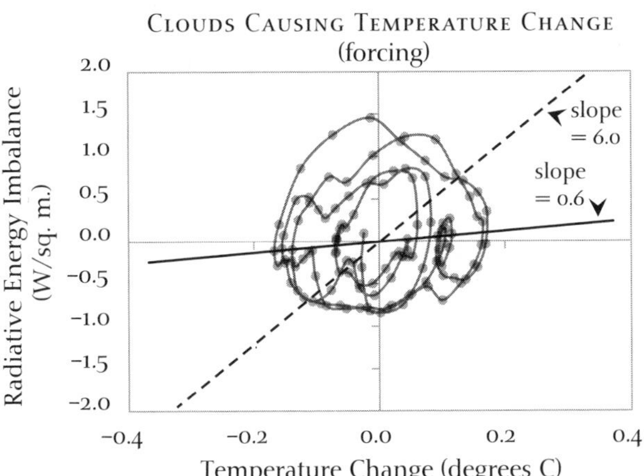

Fig. 17. A simple forcing/feedback climate model driven only by natural cloud variability demonstrates how cloud variations causing temperature change leads to a false positive feedback diagnosis.

diagnosed feedback is our first indication of why climate researchers have been fooled by Mother Nature. The reason why we cannot estimate feedback from Fig. 17 is that both the radiative forcing and the radiative feedback are mixed in together. In fact, the forcing totally obscures the feedback.

In order to estimate feedback in this case, we would first need to remove the random forcing we imposed on the model from the model output, and then plot the data and fit a line to them. In fact, this is exactly how P. M. Forster and K. E. Taylor diagnosed the feedbacks operating in all the IPCC climate models.[9] They removed the anthropogenic radiative forcing imposed on the models as part of the model global warming experiments, then plotted the remaining, feedback part of the radiative fluctuations against the temperature changes. They then fit lines to the data, and the slopes of those lines provided pretty accurate estimates of the net solar and infrared feedbacks occurring in those models in response to the anthropogenic forcings.

But while this forcing-removal procedure is possible with climate models, we cannot remove natural forcings from satellite measurements of the real climate system because we don't know what those natural forcings are! Except in the case of a single, rare event like the eruption of Mt. Pinatubo in 1991, what we see in the real climate system are natural, quasi-random, chaotic cloud variations.

Fortunately, it turns out that these natural, quasi-random forcings have a characteristic signature. The bottom panel of Fig. 17 shows looping features, which I call "radiative forcing spirals." It turns out that these radiative forcing spirals are a necessary consequence of any realistic time-varying radiative forcing of global average temperature. Since the temperature response lags the radiative forcing, the modeled climate system is in a perpetual state of energy imbalance, as represented by the spiral patterns. Through experimentation with the simple climate model, I found this spiral or looping behavior to be independent of whether the cloud forcing is random or cyclical, how deep the model ocean is assumed to be, or whether positive or negative feedback is operating.

True energy balance and temperature equilibrium would occur only at the origin of the graph, where the horizontal and vertical axes meet. That is the only point at which there is no energy imbalance and no temperature response. But true energy balance never occurs, except briefly and accidentally as the system travels from positive to negative states of imbalance and then back again.

If you attempt to do a statistical fit of a line to the data to estimate feedback when there are only natural cloud variations occurring, then that line will always be nearly horizontal, that is, a line with zero slope. As I mentioned before, a horizontal feedback line corresponds to a borderline unstable climate system. What this means from a practical standpoint is that when one tries to estimate feedbacks from satellite data, natural fluctuations in cloud cover of the Earth will cause a temperature response that gives the illusion of a borderline unstable climate system – even if the true feedback operating in the climate system is strongly negative.

As far as I have been able to determine, this fundamental problem has never been addressed before. Climate researchers have simply disregarded the effect that natural cloud variations in the climate system have on their interpretation of feedbacks. In their attempts to determine how temperature variations cause clouds to change (feedback), they have ignored how cloud variations cause temperatures to change (forcing).

The True Signature of Feedback

I realize that the evidence I have presented so far might be less than convincing, as it leaves some questions unanswered. For instance, if radiative forcing due to natural cloud variability causes the feedback line to remain horizontal, how did anyone ever estimate feedbacks from satellite data, such as the data previously shown in Fig. 14? In other words, why do the satellite data in Fig. 14 have a slope to them, while the data output from my simple model do not?

The answer is related to something I have kept from you until now. Remember when I said that the only way to force global warming or cooling was from a radiative imbalance due to either solar or infrared radiation? Well, I lied.

In dealing with climate variability on time scales of a few months or less, there are two kinds of forcing that can cause a temperature change: radiative and nonradiative. It is really the total *heat content* of the climate system that can only be radiatively forced. But temperature and heat content in the climate system are not the same thing. This is because the system is made up of three very different kinds of heat reservoirs: land, ocean, and atmosphere.

Most of the sunlight that is absorbed by the Earth is absorbed by the land or the ocean. This heat energy is then transferred to the atmosphere, but at a rate that fluctuates somewhat over time. When an extra amount of heat is given up by the ocean to the atmosphere, there will be only a small temperature decrease in the ocean, but a much larger temperature rise in the atmosphere. This is an example of nonradiative forcing of temperature. Since the temperature changes in the ocean and the atmosphere were not caused by a radiative imbalance of the Earth but instead by an exchange of energy between the ocean and the atmosphere, it is referred to as nonradiative forcing of temperature change.

Partly because 70 percent of the Earth is covered by ocean, the most common way for these nonradiative forcing events to occur is with variations in the rate of evaporation from the ocean surface. These variations can be driven either by a change in the average wind speed blowing over the ocean surface, by a change in the temperature difference between the ocean and the lower atmosphere, or by a change in the humidity of the air flowing over the ocean.

The most prominent example of this kind of event in nature is the Madden-Julian Oscillation, or MJO, which occurs sporadically in the tropical Pacific, lasting for about one or two months.[10] Wind speeds and evaporation rates over the ocean increase above normal, and the ocean surface cools below normal. This loss of extra heat through enhanced evaporation is the same mechanism that cools your skin when it is wet: your body gives up extra heat in order to evaporate the water from your skin, and that "latent heat" energy is stored in the water vapor.

In an MJO event, the increased evaporation causes greater rainfall in what is now an unusually humid tropical atmosphere.[11] When that extra rain forms in shower and thunderstorm clouds, it releases the excess heat that was lost by the ocean during enhanced evaporation. This release of latent heat raises the temperature of the atmosphere considerably—by much more than the ocean temperature fell. During these enhanced atmospheric heating events, the global average atmospheric temperature can rise by 1 deg. F or more in a matter of weeks. That is equivalent to fifty years of global warming packed into a single month.

This also means that on time scales of one or two months, an unusually warm atmosphere can occur simultaneously with an unusually cool ocean surface, and vice versa. This complicates the identification and measurement of feedbacks, because by definition feedbacks are caused by a temperature change. So, which temperature do we use: the atmospheric temperature, the surface temperature, or some combination?

Hurricanes are a particularly extreme, although localized, example of this kind of nonradiative forcing of temperature change. High up in the atmosphere, the core of a major hurricane can be as much as 10 deg. C (18 deg. F) warmer than the air surrounding the storm.[12] This extra heating comes from the condensation and precipitation of the extra water vapor gathered from the surrounding ocean by the strong winds circling the hurricane. In contrast to the huge atmospheric temperature increase in the hurricane's core, the cooling of the surrounding ocean might amount to only 1 deg. C (1.8 deg. F).

Now that we have a better understanding of nonradiative forcing events, let's run our model with nonradiative forcing to see what kind of behavior emerges. As in the radiative forcing case shown in Fig. 17, we will assume that the forcing varies randomly with time. I have adjusted the random fluctuations in nonradiative forcing to be somewhat stronger and more frequent to match the kind of disturbances that occur in nature, but this does not affect the conclusions we will make.

The resulting temperature behavior in the top panel of Fig. 18 looks pretty similar to the previous model run where we used only radiative forcing. In fact, the temperature doesn't really "care" whether the forcing is radiative or nonradiative because the temperature change will be the same either way.

But our satellite measurements of radiative imbalance do "care" whether the forcing is radiative or nonradiative. If the forcing is radiative, then the satellite will measure a combination of forcing and feedback. If the forcing is nonradiative, then the only radiative change the satellite will see is that due to feedback. Significantly, when we plot the model output on a graph, instead of seeing radiative forcing spirals like those in Fig. 17, we see in the bottom panel of Fig. 18 that the data fall neatly on a straight line, the slope of which exactly matches the feedback I specified in the model run.

Why do the plots in Figs. 17 and 18 look so different, with one producing spirals and the other producing a straight line? It's because in Fig. 18 there was no radiative forcing (which produces spirals) to obscure the radiative feedback. Put differently, there was no "cause" (radiative forcing) to obscure the "effect" (radiative feedback).

These two different model simulations of temperature variability – one with radiative forcing representing natural cloud fluctuations, the other with nonradiative forcing representing variations in the transfer of heat between the ocean and the atmosphere – suggest that feedbacks can be estimated from satellite data only in response to nonradiative forcing events, not to radiative forcing events. We will see that this has profound implications for estimating feedback from satellite data. While this result has been published, it appears that no one has yet understood its implications.[13]

As you may already have guessed, in the real world there is a mixture of radiative and nonradiative forcings, in varying proportions, on a continuous basis. So now let's run the simple model with approximately equal amounts of radiative and nonradiative forcing. The results, shown in Fig. 19, reveal a blending together of the two extreme cases.

Significantly, the cloud of points in the bottom panel of Fig. 19 now more closely resembles the satellite data shown in Fig. 15. The nonradiative forcing causes a tendency for the data to slope upward to the right, what I call "feedback stripes," while the radiative forcing causes scatter in the data that pushes the slope of the line toward zero.

At this point, the important thing to notice is that the solid

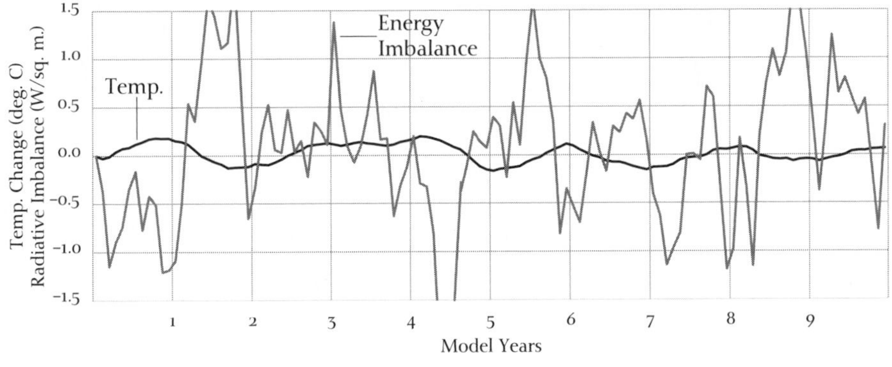

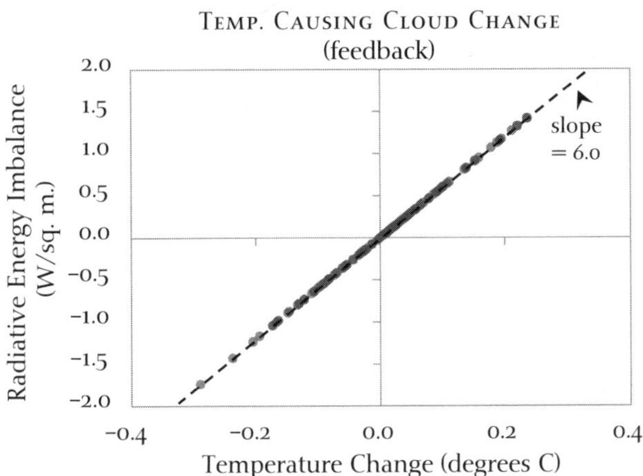

Fig. 18. A simple forcing/feedback climate model driven only by variations in evaporation and precipitation results in model output that allows an accurate diagnosis of feedback.

line fit to the data in Fig. 19, which represents the traditional method of diagnosing feedbacks from satellite data, has a slope that is significantly less than the true feedback (dashed line) that was specified in the model run. What this means for the diagnosis of feedbacks from satellite data is that when there is a mixture of radiative and nonradiative forcings of temperature occurring, *natural cloud fluctuations in the climate system will cause a bias in the diagnosed feedback in the direction of positive feedback, thus giving the illusion of an overly sensitive climate system.*

If the real climate system looks sensitive to climate modelers, they will build their models to be sensitive also. This then causes the models to produce large amounts of global warming in response to anthropogenic greenhouse gas emissions. Forcing and feedback (cause and effect) have been confused in previous interpretations of natural cloud and temperature changes. As a result, climate researchers analyzing the data have been fooled into believing that the climate system is much more sensitive than it really is.

This also provides an explanation for something I noted earlier: feedback estimates diagnosed by other researchers have been quite variable. This is because, depending on the relative proportions of radiative versus nonradiative forcing present, the contamination of the true feedback signal will vary considerably. This will cause different feedback slopes to be diagnosed, depending on what historical period of satellite data is analyzed, even if the feedback never changes. The variability in slope then leads to great uncertainty regarding what the true feedbacks are that operate in the climate system.

But we see that it is probably not the feedbacks themselves that have been varying. That would be quite worrisome from the standpoint of trying to predict climate change. Instead, different proportions of radiative versus nonradiative forcing have led to various levels of contamination of the feedback signal. This variable amount of radiative forcing is presumably being caused by all kinds of internal, chaotic natural variability in clouds. Other researchers have noted the presence of such natural variability in the climate system on a year-to-year basis, but they did not fully appreciate its impact on our understanding of how the climate system works.

Finally, it should be pointed out that observational estimates of negative feedback in the climate system are not new. For instance, one detailed analysis of the climate system's response to the 1991 eruption of Mt. Pinatubo has yielded a feedback estimate of 4.5 W m^{-2} K^{-1}, which would correspond to only 0.8 deg. C (1.5 deg. F) of warming for a doubling of atmospheric carbon dioxide.[14]

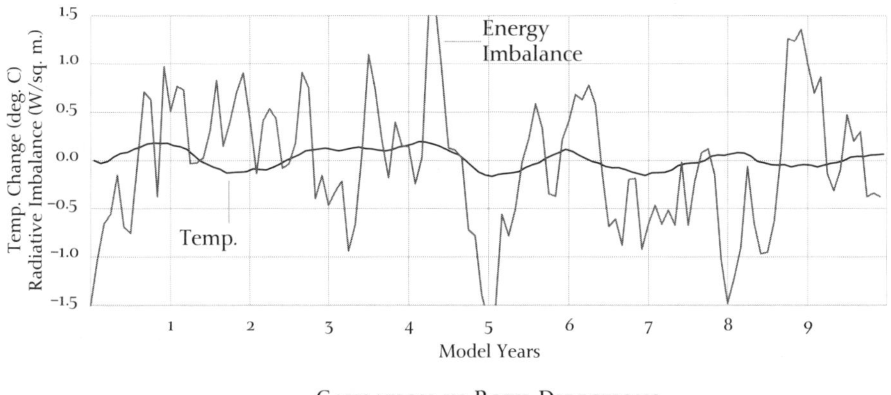

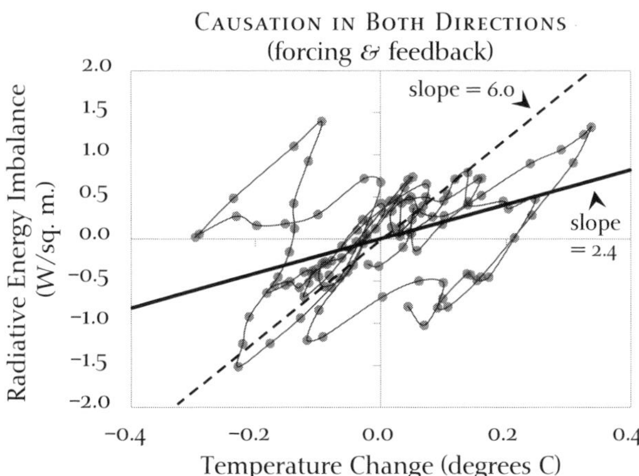

Fig. 19. A simple forcing/feedback climate model driven by both radiative and nonradiative forcings like those that occur in the real world. The radiative forcing tends to obscure the signature of true feedback, represented here by the dashed line, which results in an underestimate (solid line) of feedback.

The IPCC is aware of these negative feedback estimates; it's just that they don't believe them.

Evidence from the IPCC Models

Since you still might be unconvinced of my interpretation from this extremely simple climate model, let's take a look at the output from the state-of-the-art climate models that the IPCC tracks. Do these models also show radiative forcing spirals and feedback stripes? And if so, do the feedback stripes match the long-term feedback that others have diagnosed from these models? If they do, then we have more justification to use the feedback stripes seen in satellite data to find out how sensitive the real climate system is.

Earlier I noted that feedbacks in climate models are not diagnosed in quite the same way as in the satellite data. Climate models have the advantage that they can be run for 100 years or more of simulated time. During these runs they have large radiative imbalances representing increasing atmospheric carbon dioxide imposed upon them, which results in a large amount of global warming in the model. Since the radiative forcing (energy imbalance) that was imposed on the model is known exactly, it can be removed from the model output so that only the radiative feedback signature remains. Plots of the data can then be made like those in Figs. 17, 18, and 19, and the data usually fall rather neatly along a line, the slope of which corresponds to the total feedback operating in the model.[15]

In the case of diagnosing feedbacks from satellite measurements of natural climate fluctuations, we are not so fortunate. There is only a relatively short period of record – typically less than ten years – so the major signal in the satellite data is not long-term warming created by a known amount of radiative forcing, but year-to-year natural variability. And since we have no idea what kinds of internally generated radiative forcings might have caused that temperature variability, there is no way to accurately remove them to estimate the remaining feedback signal. Our only hope at this point is that the radiative forcing spirals and feedback stripes also exist in both the satellite and the model data, and that these features can be explained with the simple model analysis presented above. We would then have more justification for saying that the feedback stripes seen in the satellite data really do reflect the long-term feedbacks that are operating with global warming.

Well, it turns out that in every one of the eighteen IPCC models whose output we analyzed there was clear evidence of the radiative forcing spirals. One example is shown in Fig. 20, where I have plotted yearly global average variations in surface temperature and radiative balance from the Geophysical Fluid Dynamics Laboratory (GFDL) climate model.

So the climate models themselves show chaotic cloud fluctuations causing year-to-year changes in global average temperature. As far as I have been able to determine, this is the only possible explanation for such patterns in the data. This is because there are only two kinds of forcing of temperature change: radiative and nonradiative. There is no third category. And, as demonstrated with my simple model analysis above, only radiative forcing causes spirals, while only nonradiative forcing causes feedback stripes.

You might think that the possibility of clouds causing temperature change would be uncontroversial, but a reviewer of one of

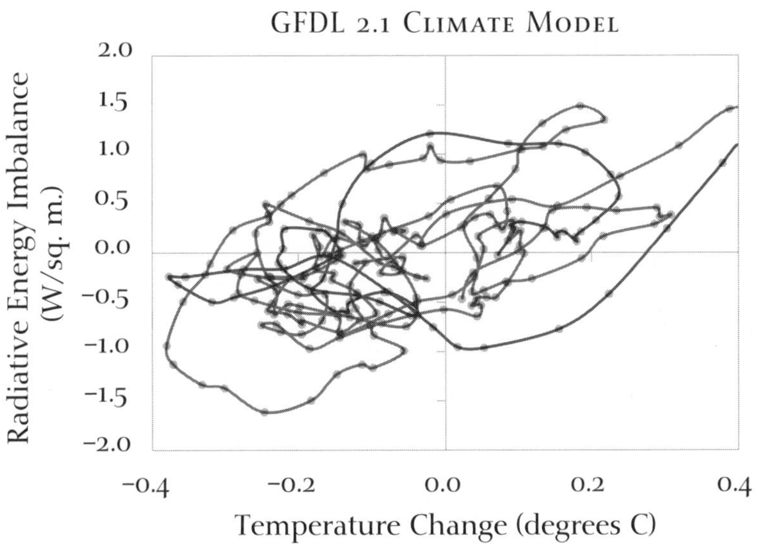

Fig. 20. The GFDL CM2.1 climate model tracked by the IPCC shows evidence of internal radiative forcing fluctuations, probably due to clouds. Running yearly averages are plotted every month from sixty years of model output.

my articles protested that clouds cannot be a cause of climate variability. Yet every one of the IPCC climate models shows evidence that they are—at least for year-to-year climate variations. And this begs a more intriguing question, which I will address in the next chapter: Can clouds cause longer-term climate change, such as global warming? The climate modelers cannot plausibly claim that clouds do not cause climate variability when their own models routinely show otherwise.

While there was abundant evidence of radiative forcing spirals in all of the IPCC models we analyzed, the search for feedback stripes proved more difficult. Feedback stripes require nonradiative forcing events, and it is well known by the climate modeling community that the models do not exhibit such strong Madden-Julian Oscillations as those that occur in nature.[16] As can be seen in Fig. 21, I found only a handful of models with fairly obvious feedback stripes, and these occurred only in the infrared component of the models' energy balance, not in the reflected solar part.

What is significant about these stripes is that their slopes are aligned with the models' long-term feedbacks as diagnosed by Forster and Taylor. In fact, I have drawn the dashed lines in Fig. 21 to correspond to those long-term feedbacks, and not to the linear striations seen in the data. That the dashed lines are roughly parallel to the linear striations is therefore evidence that the short-term and long-term feedbacks in the models are substantially the same.

This is important from the standpoint of trying to determine long-term feedbacks in the climate system from relatively short-term satellite datasets. If the climate models suggest that the short-term and long-term feedbacks in the models are the same, then it would be difficult for anyone to claim that short-term feedbacks diagnosed from satellite data do not apply to the global warming problem.

Now let's go back and apply what we have learned to the satellite data shown in Fig. 15.

* * *

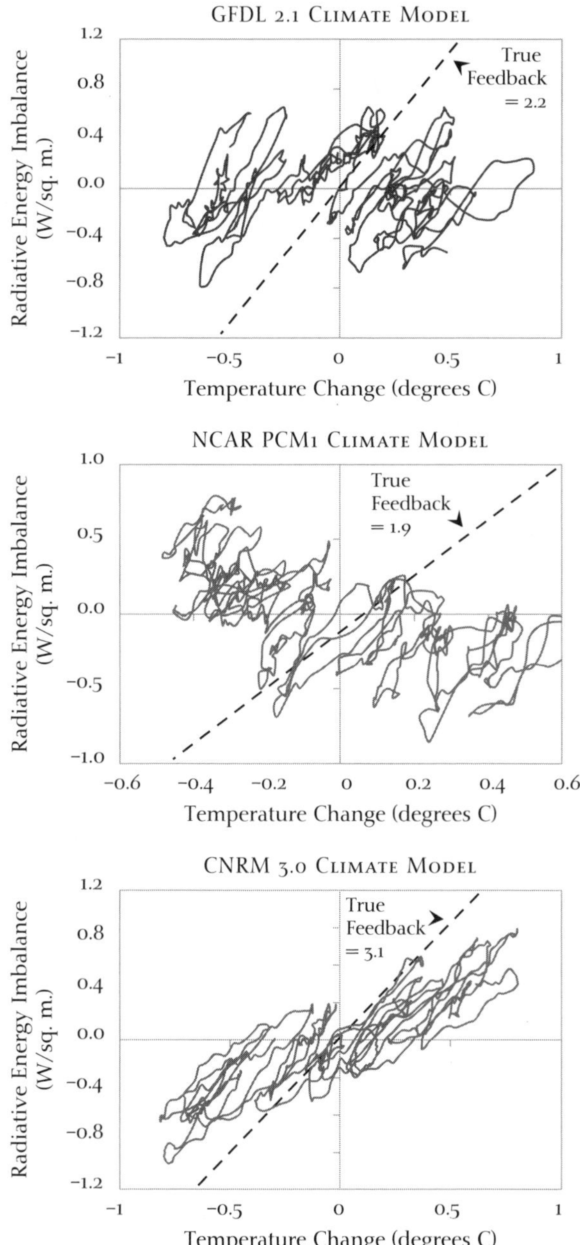

Fig. 21. Infrared feedback diagrams for three IPCC climate models all show feedback stripes that match the model's long-term feedback in response to atmospheric CO_2 increases (dashed lines).

SATELLITE ESTIMATES OF FEEDBACK (THE RIGHT WAY)

We now have a way to explain some of the linear features we saw in the satellite data shown in Fig. 15, which I have replotted in Fig. 22. Based on the preceding analysis, these striations represent periods of time when radiative forcing due to cloud fluctuations was weak, but nonradiative forcing of temperature change was strong. These temperature changes resulted in a radiative feedback response in the atmosphere, which then produced the feedback stripes.

Significantly, the slope of these lines corresponds to strongly negative feedback: around 6 watts per square meter per degree C of temperature change. To illustrate how strong this negative feedback is, it would correspond to little more than 0.5 deg. C of warming from a doubling of atmospheric CO_2. This is well below even the lower limit of future warming, 1.5 deg. C, that the IPCC is 90 percent certain will occur.

But if we instead mistakenly assume there is no radiative forcing by clouds going on in Fig. 22, we would then fit the solid line

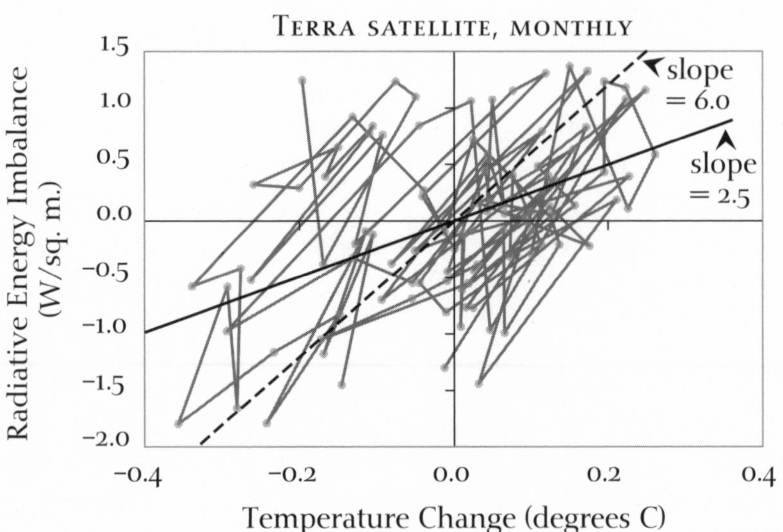

Fig. 22. As in Fig. 15a.

to the data, the slope of which corresponds to weakly positive feedback, and about 1.5 deg. C of warming for a doubling of CO_2. This is a considerable difference in diagnosed climate sensitivity. If global warming by 2100 is below 1 deg. C, then manmade global warming is for all practical purposes a false alarm. By then we will probably have new energy technologies that greatly reduced our dependence on fossil fuels. I might even have to reconsider James Hansen's claim that the Greenland ice sheet will eventually disappear.

The difference between these two interpretations is the difference between an uncritical analysis of the data, which incorrectly assumes that only feedback is being measured in these graphs, and a more thorough analysis where we account for both forcing and feedback. Again I emphasize that the problem stems from previous researchers not accounting for natural cloud variations in the climate system.

Garbage In, Garbage Out

Even though the analysis above clearly reveals the effect of clouds on the climate system, the IPCC has ignored any such radiative forcing generated internal to the climate system as a potential source of climate change. They are concerned only with "external" sources of forcing, such as manmade pollution, volcanoes, or tiny changes in the output of the sun. When the IPCC reports mention "radiative forcing," these are the sources they have in mind, not natural cloud fluctuations.

I believe that this neglect of natural cloud fluctuations has been the Achilles' heel of the so-called scientific consensus on global warming. By ignoring natural variability in clouds, researchers have reached the conclusion that the climate system is very sensitive to mankind's pollution. This, they argue, means that no natural source of climate change is needed to explain global warming since humanity's greenhouse gas emissions are sufficient to cause the warming we have seen over the last 50 to 100 years.

But this is circular reasoning, or a tautology. The IPCC merely ends up concluding what they assumed to begin with. By ignoring natural climate variability, they "prove" that there is no need for natural climate variability to explain global warming. They can even claim that their explanation is self-consistent–but then, that is true of any circular argument, isn't it?

What is somewhat puzzling about the IPCC's neglect of natural radiative forcing due to cloud variations in diagnosing feedbacks is that it was IPCC scientists themselves who first recognized that radiative forcing can corrupt estimates of feedback, as I mentioned earlier. The rather large radiative forcing caused by the major eruption of the Mt. Pinatubo volcano in 1991 was estimated and removed from satellite data by IPCC scientists so that the feedbacks upon that forcing could also be estimated.[17]

While I question whether their estimate of the Pinatubo forcing was known well enough to diagnose feedback accurately, all I have done is apply the same concepts that these IPCC researchers have published to an analysis of the satellite data. The only difference is that I have allowed for the presence of natural cloud variability causing temperature changes–something that every one of the IPCC models shows evidence of! And we have found that, depending on how much natural cloud variability was occurring, a different feedback (line slope) will be diagnosed from the satellite data, giving the false impression that feedbacks in the climate system are not only variable but also positive, which results in the illusion of a sensitive climate system.

What It All Means

Don't be discouraged if you don't understand these plots of data and my interpretation of them. All this has just been a quantitative way of demonstrating that climate researchers have not accounted for clouds causing temperature change (forcing) when trying to estimate how much temperature change causes clouds to change (feedback).

In simple terms, they have mixed up cause and effect when analyzing cloud and temperature variations.

As a result of this mix-up, the illusion of a sensitive climate system (positive feedbacks) emerges from their analysis. Thinking that the climate system is very sensitive, the climate modelers then built overly sensitive models that produce too much global warming.

Or, to illustrate the issue another way, let's return to the question I had when I got involved in this line of research. When researchers have observed clouds decreasing with warming, they have claimed that this is evidence of positive feedback—a sensitive climate system. They have explained that the warming causes the clouds to decrease, which then amplifies the warming.

But how did the researchers know that the warmer temperatures caused the clouds to decrease, rather than the reverse? In other words, how did they know they weren't mixing up cause and effect? It turns out they didn't know. We now have peer-reviewed and published evidence of decreases in cloud cover causing warmer temperatures, yet it has gone virtually unnoticed.

I believe that this misinterpretation of how clouds really behave in the climate system helps explain why the scientific consensus is so sure that mankind is causing global warming. By confusing natural variability in clouds with positive feedback, researchers have been led to believe that the climate system is very sensitive. This, in turn, has led them to conclude that the small amount of forcing from humanity's greenhouse gas emissions is being amplified enough to explain most of the global warming that we have seen in the last fifty years or more. They claim that no natural explanation is needed for warming—that humanity's pollution is sufficient.

By ignoring natural variations, they have concluded that they can ignore natural variations. The circular nature of their reasoning has not occurred to them.

Furthermore, natural variability in clouds probably also explains why climate sensitivity estimates have been so variable when previous researchers have diagnosed feedbacks from satellite data. Depending on how much natural cloud variability was occurring when the satellites made their observations, a wide variety of feedback (climate sensitivity) estimates would result–some bordering on a catastrophically sensitive climate system. And as long as the IPCC can claim that feedbacks in the real climate system are very uncertain, they can perpetuate their warnings that disastrous global warming cannot be ruled out. They tell us that the sensitivity of the climate system is high, but just how high isn't really known for sure. Therefore, we must prepare for catastrophic warming, just in case.

One detail that I did not discuss in this chapter is how the infrared and solar parts of feedback behaved during the period for which we have satellite data. It turns out that the negative feedback seen by the satellites was entirely in the reflected solar component, which is most likely due to low clouds. The infrared portion of the feedback supported positive water vapor feedback, which is consistent with feedback estimates from other researchers. But it is the total feedback–solar plus infrared–that determines climate sensitivity. If negative feedbacks outweigh positive feedbacks, then the net feedback is still negative.

Even the IPCC recognized the uncertainty associated with reflected solar feedback from low clouds in their 2007 report when they concluded: "Cloud feedbacks are the primary source of inter-model differences in equilibrium climate sensitivity, with low cloud being the largest contributor."

Taken together, all this evidence indicates that the climate models are too sensitive, which is why they predict so much global warming for the future. In contrast, the satellite evidence indicates that the climate system is quite insensitive, which means that it doesn't really care how big your carbon footprint is. Rather than 1.5 to 6 deg. C (or more) of warming as predicted by the IPCC, a careful examination of the satellite data suggests that manmade warming due to a doubling of atmospheric carbon dioxide could be less than 1 deg. C (1.8 deg. F)–possibly much less.

But if anthropogenic global warming will amount to less than 1 deg. C for a doubling of atmospheric CO_2, why have we already seen at least 0.7 deg. C (1.3 deg. F) of warming? In the next chapter I will reveal evidence that the answer is fundamentally the same as the reason why scientists have been fooled into believing that the climate system is sensitive: they have ignored natural variations in clouds.

Chapter 6 · Global Warming: Satellite Evidence for an Alternative Explanation

If the climate system is relatively insensitive to our greenhouse gas emissions, then what has caused the warming observed over the last 100 years? New NASA satellite measurements reveal that the Pacific Decadal Oscillation causes a radiative forcing of the Earth—probably due to a change in low cloud cover of the oceans—that is sufficient to explain most of the temperature variability during the twentieth century, including 75 percent of the global warming trend.

If I am correct in asserting that total feedbacks in the climate system are negative and that the climate system is relatively insensitive to our greenhouse gas emissions, this still leaves an important question unanswered: What has caused the warming that we have experienced in the last 100 years?

Can Mother Nature Cause Global Warming?

The IPCC claims that global warming or global cooling will occur only in response to an energy imbalance imposed on the system "externally." That is, the climate system keeps itself in an overall state of energy balance unless a volcano, the sun, or mankind comes along and upsets that balance. This is a fundamental assumption that is seldom mentioned although it is central to the IPCC's claim that humans are the main driver of the climate system. But it is only an assumption. And, it is contradicted by the last 2,000 years of temperature variations, shown in Fig. 1, Chapter 1. Global warming and cooling are not the exception but rather the rule over the last twenty centuries. And it is pretty obvious that it is nature itself that is in control of these changes.

As we saw in the previous chapter, the climate system can generate an energy imbalance *all by itself*. In fact, the satellite evidence shows that it does so every year, most likely through natural variations in cloud cover. We even saw evidence of cloud variations causing year-to-year temperature variations in the IPCC models. I call this natural source of climate variability chaos because it is complex, little understood, and largely unpredictable. The clear evidence of this was shown to be "radiative forcing spirals," which occur both in satellite observations and in data output from the IPCC climate models. As far as I have been able to determine, there is no other physical explanation for this behavior.

If natural cloud fluctuations can cause year-to-year temperature variations, then why not changes over 10 years? Or 50 years? Or 100 years? Do chaotic cloud variations occur on the longer time scales involved in global warming and climate change? In the first chapter, I showed that the temperature proxy data for the last 2,000 years are dominated by large swings in temperature, quite similar to the warming we have seen in the last 100 years or so.

But the IPCC doesn't seem to want to address any evidence for natural causes of global warming and climate change. Apparently, doing so might marginalize the role of humans in climate change. This is why the hockey stick became so famous: it virtually eliminated natural climate variability. I find this position to be astoundingly naïve, if not scientifically corrupt. El Niño, La Niña, the Pacific Decadal Oscillation, and other natural modes of climate variability are a few examples of chaos in the climate system. The IPCC knows they exist. These climate modes have preferred time scales, a degree of regularity that has caused them to be identified and named in the first place. Yet the IPCC assumes that these cause only regional changes in weather patterns, not changes in global average cloud cover. This is such a consequential assumption that you would think they have data to back it up. They have not provided any such data.

And what about other sources of year-to-year, or even decade-to-decade, climate variability that do not have names? While these fluctuations might never be very well understood or predicted, they still have the potential to cause small changes in global average cloud cover. If such natural fluctuations persist long enough, they also have the potential to cause global warming or cooling. Just because we don't understand them doesn't mean we can assume they don't exist.

It has always been my opinion that the possibility of natural, internally generated climate change is ignored by the climate modelers because most of them are not sufficiently well versed in meteorology. They assume that the climate system magically stays the same indefinitely, and that the only way for it to change is by being *forced* to change through some influence external to weather, such as a volcano, or a fluctuation in the sun's output, or manmade pollution.

But we meteorologists understand that the processes controlling clouds, "nature's sunshade," are myriad and complex. I have found that most meteorologists readily accept the possibility of natural climate change. They have no problem admitting that there are things we still do not understand about the climate system. For instance, in my experience a majority of TV meteorologists do not believe the claim that global warming is manmade. According to Dr. John Christy, my associate and Alabama's state climatologist, the same sentiment is also held by most of the country's state climatologists.

As I mentioned previously, some IPCC scientists have chided TV meteorologists for second-guessing them on the issue of global warming. They argue that meteorologists deal with weather, not climate, and therefore should not question the judgment of climate experts when it comes to global warming. But I would turn this around: I contend that climate variability cannot be understood without first understanding the complexities of weather. After all, climate is average weather, and if you don't understand what controls variations in weather then you won't be able to understand all the potential sources of climate change.

Our 2008 paper demonstrated how something as simple as daily random fluctuations in cloud cover can cause substantial temperature trends over ten years. So, what might yearly, 10-year, or 30-year chaotic fluctuations in cloudiness do? Maybe the Medieval Warm Period and the Little Ice Age are examples of chaos generated by the climate system itself.

I have found that the public is also perfectly comfortable with the idea of natural climate cycles that we do not yet understand. The public accepts natural climate variability as a fact of life, supported by historical evidence. People see with their own eyes the power and complexities of weather on a daily basis. If such things can be appreciated by the public, why do the climate experts find them so difficult to accept?

It would take natural variations of little more than 1 percent in global average cloud cover to explain most of the climate change seen in the last 2,000 years, yet our ability to measure such small changes has existed for only the last ten years. Without any evidence available to prove them wrong, the IPCC can simply assert that this does not happen. How convenient. Given the basic nature of scientific inquiry, I find the IPCC's resistance to the idea of natural climate change very peculiar. Science always seeks alternative explanations for observed phenomena—except, apparently, when it comes to global warming. But then, as I have mentioned before, the IPCC was formed for largely political reasons, not scientific.

To be fair, the IPCC's failure to investigate natural, internal mechanisms of climate change more thoroughly is partly the result of not having very much data to investigate. To actually prove that Mother Nature has caused global warming, one would need many decades of highly accurate satellite measurements of the entire Earth. It would be necessary to document that the Earth does go through extended periods of energy imbalance, which would then cause extended periods of warming or cooling. But we presently have less than ten years of such high-quality data, and even these measurements are not accurate enough to determine the imbalance due to the extra carbon dioxide in the atmosphere. That energy imbalance must instead be computed theoretically.

It has been very convenient for the more alarmist climate scientists and politicians that global data for a thorough investigation of natural sources of global warming do not exist. This allows them to claim that all we need to know in order to explain global warming is that CO_2 is a greenhouse gas and that CO_2 concentrations in the atmosphere are increasing. No other explanation is necessary, they insist.

This is an example of what is called an anthropic bias. We think we are the ones changing the climate system because the only climate forcing mechanism we understand well is the one we also happen to produce: carbon dioxide. The belief in catastrophic global warming was even described by the author Michael Crichton in religious terms, with strong parallels between traditional biblical practices and green practices.

Perhaps Mother Nature has been trying to tell us, through our satellite measurements, that natural climate change is going on, but we just haven't been listening. Consider, for example, the gradual melting of sea ice that has been observed in the Arctic Ocean over the last thirty years. The depletion of summer sea ice is well documented by satellite observations, and it is one of the poster children for global warming. Climate model predictions of more sea ice melt have even led to the listing of the polar bear as "threatened" under the Endangered Species Act—although there is little evidence that the total population of polar bears has even been affected by changes in sea ice conditions.[1] Apparently, it doesn't hurt that polar bear cubs are irresistibly cute. I wonder if the same thing would have happened if it were the Arctic sea slug at risk of extinction from disappearing sea ice.

What if this warming in the Arctic is mostly natural? As I mentioned in Chapter 1, thermometer measurements circling the Arctic Ocean have shown that it was nearly as warm there in the 1930s as it is today. The news media reported Arctic sea ice to be receding then, too. The Northwest Passage was probably open in the early 1940s, just as it was again in 2007. We really do not know whether the recent "record" low in sea ice extent, in the fall of 2007, was any lower than it was in the late 1930s or early 1940s. We have had accurate satellite observations of the Arctic Ocean only since the mid-1970s. Just by coincidence, the Pacific Decadal Oscillation went into its positive, warming phase around the same time—so all of our satellite measurements of sea ice have been made during the warming phase of this natural climate cycle.

Surely this is known to the IPCC. Their latest report even gives a cursory nod to such natural modes of internal climate variability, but not in the context of global warming. The possibility that these natural modes of climate variability might explain a good part of climate variability in the last 100 years has been largely ignored. The IPCC has not acknowledged that changes in the general circulation of the ocean-atmosphere system might cause associated changes in cloud cover, which then would change Earth's energy balance over long periods of time.

The Pacific Decadal Oscillation as an Agent of Climate Change

As discussed in Chapter 1, the Pacific Decadal Oscillation changes phase every thirty years or so. The positive and negative phases of the PDO can be thought of as representing two different preferred types of weather patterns that become established over the North Pacific Ocean and then persist for decades.

The upper panel in Fig. 23 shows how the PDO index has varied over the last century. It reveals three main features, which happen to coincide with three periods of changing temperatures shown in the bottom panel of Fig. 23.

The PDO was in a persistent positive phase during the period between 1920 and 1940, then a negative phase through the mid-1970s, then back to a positive phase since the 1970s. Is it just a coincidence that these three features correspond to global

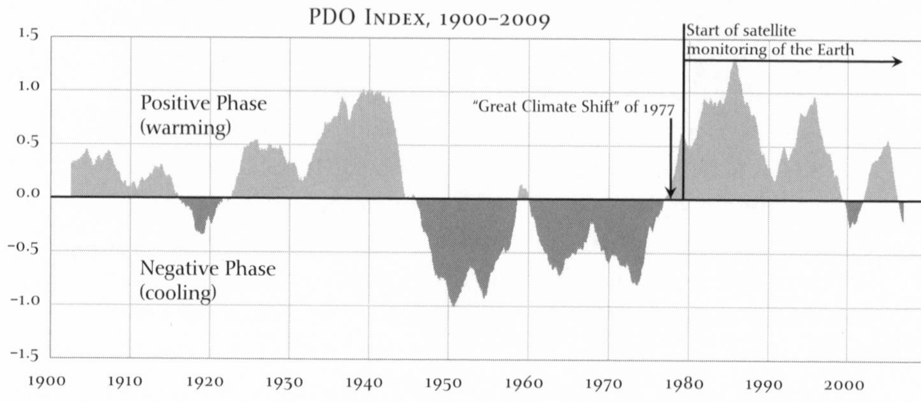

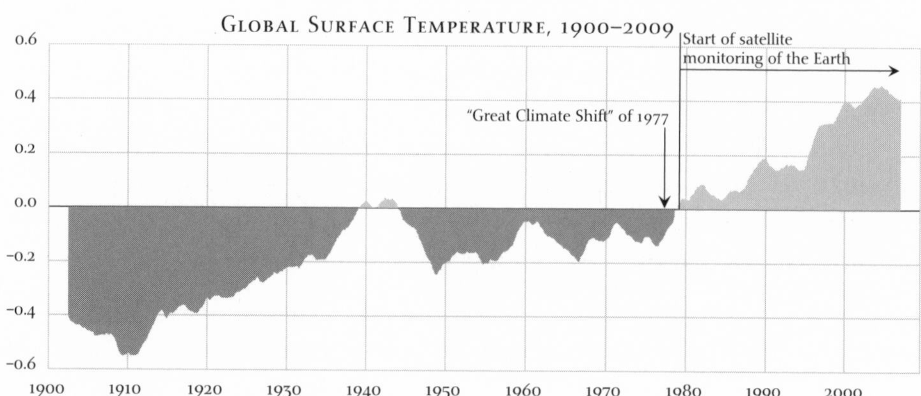

Could Natural Cloud Variations Cause Global Warming?

Fig. 23. Five-year averages of the PDO index (top panel) and globally averaged surface temperature variations (bottom panel) during the twentieth century.

temperature trends that also changed in these three periods: warming up to the 1940s, slight cooling until the 1970s, and warming since then? I don't think so.

We already know that when the PDO flipped sign in 1977, an event that has been dubbed the Great Climate Shift of 1977, Alaska suddenly got warmer and has stayed warm ever since. As was shown in Fig. 3 of Chapter 1, this is a characteristic of the warm phase of the PDO, with more warm air flowing from the south into Alaska. Then, several years later our newly operational polar-orbiting weather-monitoring satellites began to observe a gradually decreasing extent of Arctic sea ice, especially during the summer melt season.

Of course, the increasing melt-back of Arctic sea ice in the summer has been widely attributed to anthropogenic global warming. But evidence for a natural change in ocean circulation in the Arctic Ocean has also been published, and that change might be a signature of the PDO.[2] The Great Lakes reached record-low water levels in the 1930s, and again in 2007.[3] The similarities between our recent climate changes and the changes that occurred up until 1940 are inescapable.

It is thus entirely reasonable to ask: Could the shift in ocean-atmosphere circulation associated with the PDO have caused a small change in global average cloud cover? If, for instance, cloud cover was below normal during the positive phase of the PDO, this would have let more sunlight into the climate system, causing a warming trend—as was seen during 1920–1940, and then again from the late 1970s up to the present.

Again I must emphasize that the PDO index is not a temperature index, but an index of how weather patterns over the North Pacific Ocean are arranged. What I am hypothesizing is that the PDO might also cause a small fluctuation in cloud cover resulting from those circulation changes. In this case, the PDO would constitute the forcing, and as we saw in Chapter 3, the temperature response to a forcing takes time to develop.

If you understand this distinction, you are doing better than some climate experts. In early 2009 I submitted the work I am describing for publication in *Geophysical Research Letters*, and the paper was quickly rejected by a single reviewer who was very displeased that I was contradicting the IPCC. Besides, this reviewer argued, because the PDO index and temperature variations shown in Fig. 23 do not look the same, the PDO could not have caused the temperature changes.

But apparently this climate "expert" had never heated a pot of water on the stove before. As I showed in Fig. 10 of Chapter 3, whether it is something as complex as the Earth's climate system or as simple as a pot of water on the stove, the maximum temperature response comes *after* the forcing is applied. The time history of the forcing therefore does not look exactly like the time history of the temperature response. When you first turn the stove on, the water *begins* to warm. And it will keep warming as long as an energy imbalance exists. Then, if you turn the stove off, the water will cool – and it will keep cooling as long as an energy imbalance exists.

This expert's comments revealed a fundamental misunderstanding of how temperature changes are caused, and as a result my paper was rejected for publication. In fact, the editor was so annoyed he warned me not to bother changing and then resubmitting it. My results, more of which will be described below, had obviously struck a nerve. This is the sorry state of scientific peer review that can develop when scientists let their preconceived notions get in the way.

Of course, in the case of a pot of water on the stove it only takes several minutes for the temperature of the water to reach a peak after the burner is turned on. But in the case of long-term climate change, it can take decades for the temperature of the ocean to fully respond to a change in cloud cover. It takes much longer to heat up a layer of water 2,000 feet deep than it does 2 inches of water in a pot.

But since this potential relationship between the PDO and climate change is just hand waving so far, we will next use the simple climate model to quantitatively test whether forcing that looks like the PDO history in the upper panel of Fig. 23 can cause a temperature history like that seen in the bottom panel. And even if it can, is there any evidence that the PDO actually does alter the amount of cloud cover on the Earth in this manner? I will now take you through the process I went through to answer these questions.

* * *

A Simple Model of Global Warming Caused by the PDO

Climate modelers have spent decades and hundreds of millions of dollars running complex climate models on supercomputers trying to explain those global average temperature based on anthropogenic pollution. I set out to determine if I could explain those features naturally with a simple model on my desktop computer at home.

Specifically, can the PDO variations shown in Fig. 23 cause the temperature variations seen in that figure? We can investigate this possibility theoretically by again using the very simple climate model introduced in Chapter 5, the basic components of which are now shown schematically in Fig. 24. Again, the model is a single equation, which states that a change in temperature of the climate system is proportional to a forcing, minus a certain amount of feedback upon that temperature change that tries to restore the temperature to its preferred equilibrium value.

The only difference from how the model was used in Chapter 5 is that instead of varying the cloud cover randomly in time as a forcing, we will now force the model with cloud changes assumed to vary in lockstep with the variations in the PDO index since 1900. Later we will examine satellite measurements of the Earth to look for evidence that such a climate forcing mechanism actually exists in nature.

There are only four adjustable parameters in this simple model, which act like "knobs" that we can turn to make the model behave differently. The first parameter is the depth of the ocean assumed to be involved in temperature change. It is like the depth of water in the pot on the stove: one inch of water will heat up much quicker than six inches of water. Given a certain amount of forcing (energy imbalance) caused by cloud variations, the rate at which the temperature changes with time is determined by how deep an ocean layer is being warmed or cooled.

A second knob we can adjust on the model is feedback. As the

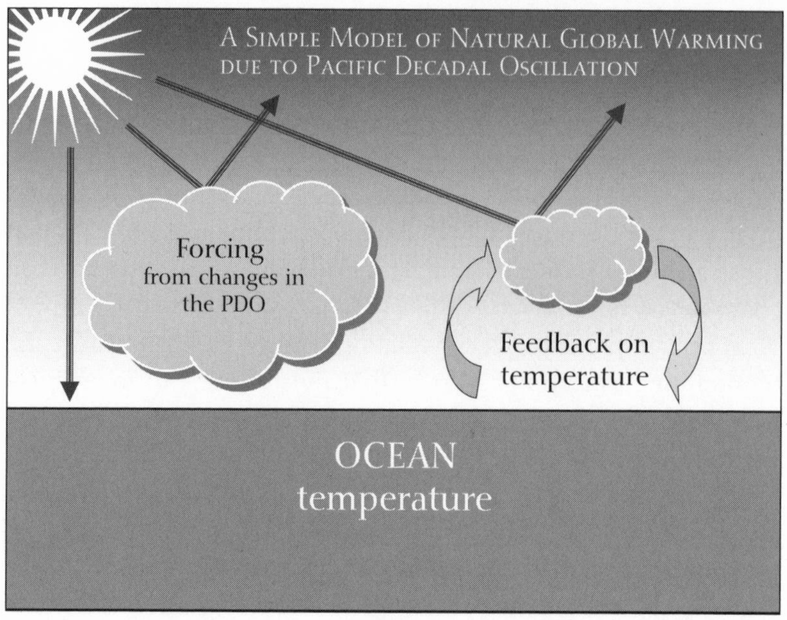

Fig. 24. A simple conceptual model of climate change driven by natural cloud variations associated with the Pacific Decadal Oscillation.

temperature goes up, does the model act to reduce the amount of warming (negative feedback), or amplify it (positive feedback)?

The third adjustable parameter is the amount of cloud change assumed to be associated with the PDO. By just how much would we need to vary global cloud cover in proportion to the PDO in order to produce temperature changes like those measured during the 1900s?

The final parameter we must specify to run the model is a starting temperature anomaly, that is, an initial temperature departure from normal. As you might recall, the simple climate model assumes that there is some "normal" climate state, a baseline temperature about which the climate varies. This baseline is important because the feedback will always be trying to "push" the temperature back to its normal value. That "push" is either strong in the case of negative feedback, or weak in the case of positive feedback. Therefore, it makes a difference whether the model run starts with a temperature above normal or below normal, and by how much.

Since we don't know how to set the four knobs on the model to cause it to produce temperature variations like those seen in Fig. 23, we will use the brute force of the computer's great speed to do 100,000 model runs, each of which has a unique combination of these four knob settings. And because spreadsheet programs like Excel aren't made to run this many experiments, I programmed the model in Fortran.

It took only a few minutes to run the 100,000 different combinations of knob settings. Out of all these model simulations, I saved the ones that came close to the observed temperature variations between 1900 and 2000. Then, I averaged all of those thousands of temperature simulations together, which produced the curve labeled "PDO" in Fig. 25. What we see is that if the computer gets to "choose" how much the clouds change with the PDO, then the PDO alone can explain 75 percent of the warming trend seen during the twentieth century. In fact, it also does a pretty good job of capturing the warming until about 1940, then the slight cooling until the 1970s, and finally the resumed warming until 2000.

If I instead use the history of anthropogenic forcings that James Hansen has compiled (the "CO_2" curve in Fig. 25), somewhat more of the warming trend can be explained, but the temperature variations in the middle of the century are not as well captured. I should note that the "warm hump" around 1940 and the slight cooling afterward have always been a thorn in the side of climate modelers. Considerable effort has been expended to try to figure out why these events occurred. Why did temperatures rise so rapidly before 1940 even though so little carbon dioxide had been emitted by then? And why did a cooling trend set in after 1940, just as humanity's carbon dioxide emissions were coming on strong? Maybe the PDO is the key.

It is also worth noting the four model "knob settings" that the computer said provided the best match to the observed temperature record. First, the model preferred an ocean depth of around

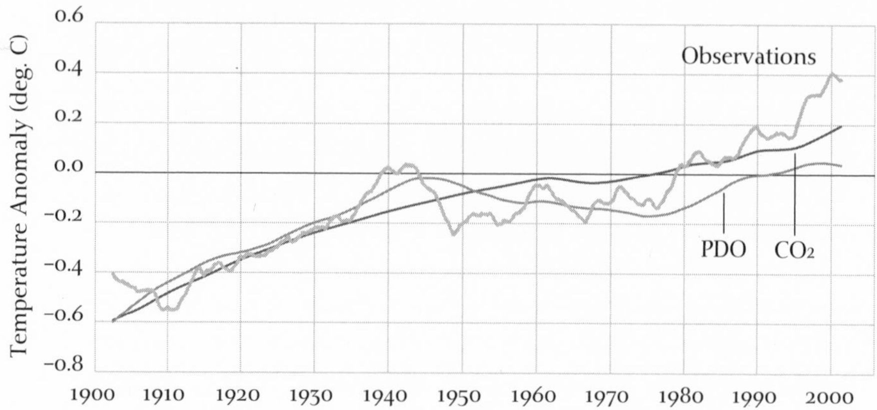

Fig. 25. A simple computerized climate model can explain most of the temperature variability in the twentieth century with natural forcings alone.

700 meters (about 2,300 feet). By coincidence, this figure actually matches the approximate depth over which warming has been observed to occur in the last fifty years,[4] which is something the model did not know beforehand.

Second, the optimum feedback chosen by the model corresponds closely to neutral feedback–neither positive nor negative–about 3.0 W m^{-2} K^{-1}. If this was the feedback operating in the real climate system, it would correspond to a global warming estimate of about 1.2 deg. C (2.2 deg. F) by around 2100. Note that this is below the lowest end of the range of warming that the IPCC claims to be 90 percent sure of for the future, 1.5 deg. C.

In fact, even if I use Hansen's forcings alone, which are dominated by humanity's aerosol and greenhouse gas pollution, the model says that the observed temperature variations during the twentieth century are still consistent with a relatively insensitive climate system.

The third parameter is the starting temperature anomaly in 1900: the model chose a temperature of about 0.6 deg. C below normal. This choice is interesting because it approximately matches what the thermometer researchers have chosen for their baseline in Fig. 23. That is, the temperature the model decided is the best transition point between "above normal" and "below normal" is the same as that chosen by the thermometer researchers.

Finally, the fourth and most important knob setting on the model is how clouds change with the PDO. Remember, this model experiment was run to see if it was even possible from a theoretical standpoint for cloud changes associated with the PDO to cause the global average temperature variations measured in the last 100 years. While the answer to that question is "yes," at this point we have no idea whether any such cloud changes actually occur in response to the PDO. The model chose to change cloud cover in proportion to the PDO with some proportionality factor. Do we have any evidence that the PDO actually causes changes in cloud cover in the needed direction, and by the needed amount? To investigate this question, I turned to the satellite data.

Satellite Evidence for the PDO Causing Global Warming

I computed the yearly average PDO index since the Terra satellite began producing data in early 2000, as well as the corresponding global oceanic averages of the radiative energy imbalance of the Earth as measured by the CERES instrument on Terra. These two pieces of evidence should provide some insight into whether the PDO is associated in any way with a change in the global energy balance.

But first, we need to make sure we do not make the same mistake that previous investigators have made: not accounting for forcing when trying to measure feedback, or in this case, not accounting for feedback when trying to estimate forcing. In other words, we need to determine how much of the satellite-measured variability in the Earth's energy balance is due to cause versus effect. Because the satellite measurements of radiative imbalance include both forcing *and* feedback, I removed the feedback based on the observed change in global atmospheric temperature during the same period of time, 6 W m^{-2} K^{-1} as indicated in Fig. 22. Note that I am not necessarily claiming that this is the feedback operating on the long time scales associated with global warming – only that it is the average feedback involved in the climate fluctuations occurring during the period when the satellite was making its measurements.

This procedure is new, but entirely consistent with previously published work. Removing forcing to estimate feedback has been done by other investigators in estimating feedbacks from the cooling caused by the 1991 eruption of Mt. Pinatubo,[5] and feedbacks from warming in the IPCC climate models caused by increasing CO_2 concentrations.[6] The only difference is that, whereas those studies removed the radiative forcing to estimate radiative feedback, I have turned this around and removed feedback to estimate the radiative forcing due to clouds. If one is a valid procedure, then the other must be as well.

Once the feedback signal is removed from the variations in total radiative balance, we are hopefully left with just the radiative forcing. The points plotted in Fig. 26 show the resulting yearly variations in the satellite-measured radiative imbalance of the Earth, plotted against the yearly variations in the PDO index. The solid line is a statistical regression fit to the data, indicating that there is indeed a change in the radiative balance of the Earth in response to the PDO.

And to my amazement, when I also plotted the computer-chosen relationship that best explained the temperature variations during the twentieth century with the simple climate model (the dashed line in Fig. 26), there was excellent agreement with the satellite-observed relationship.

Admittedly, some of this agreement could be serendipitous. But it should be remembered that nearly nine years of data went into the averages plotted in Fig. 26. That amounts to millions of satellite observations over about 2,700 days. Also, while there is some scatter in the data, what we are looking for is an average long-term relationship between the PDO and clouds, not an explanation for year-to-year temperature variability based on the

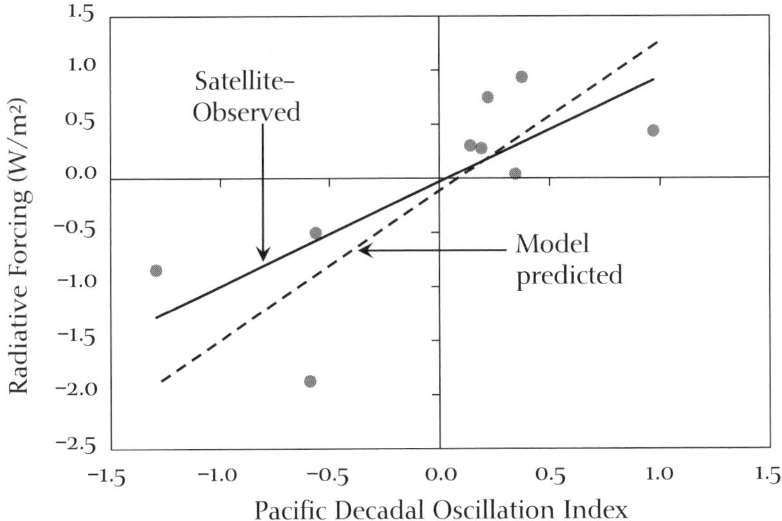

Fig. 26. Satellite observations of radiative forcing of the Earth from 2000 through 2008 suggest that the Pacific Decadal Oscillation causes natural cloud variations of a magnitude that a simple climate model indicated would be sufficient to explain most of the temperature variations during the twentieth century.

PDO. Surely there are other modes of natural climate variability occurring during different years in this period, such as El Niño and La Niña, so we would not expect a perfect one-to-one relationship between the two variables.

What It All Means

Our most accurate global satellite data, collected from 2000 through 2008, show that the Pacific Decadal Oscillation does indeed cause a change in the Earth's energy balance. Over the nine-year satellite period of record, the radiative imbalance varied over a range of at least 2.5 watts per square meter. Even though this natural source of radiative forcing is only 1 percent of the average flows of sunlight into and infrared radiation out of the climate system, the simple model analysis shows that it is sufficient to explain most of the temperature variability experienced during the twentieth century–up to 75 percent of the long-term temperature trend. This supports my original claim that a mere 1 percent change in naturally occurring processes can cause global warming or cooling.

Thus, the PDO by itself can potentially explain most of what is popularly called global warming. And while the anthropogenic explanation for global warming involves a forcing mechanism that can only be computed theoretically, the PDO forcing mechanism is actually observed by satellite. In fact, recently published research has finally begun to make this connection between the PDO and climate change, so maybe the tide is turning.[7]

This simple, natural explanation for most of the global warming experienced from 1900 to 2000 took only a desktop computer and a few days to put together. In contrast, hundreds of millions of dollars have been invested in explaining those same temperature variations with supercomputers using not just one but two manmade forcings: warming from manmade carbon dioxide and cooling from particulate pollution. This looks like a good place to apply Occam's razor, which states that it is usually better to go with a simpler explanation of some physical phenomenon than a more complicated one. Even one of our leading climate modelers, Isaac Held, has argued for less complexity and more elegance in climate modeling: "An elegant model is only as elaborate as it needs to be to capture the essence of a particular source of complexity, but is no more elaborate."[8] The IPCC has been using an expensive, complicated crowbar to try to force-fit a manmade explanation for climate change involving multiple kinds of pollution. I used a single, known mode of natural climate variability, and it fit more like a hand in a glove.

How can the IPCC leadership be so sure that global warming is manmade, when they never even investigated possible natural sources of warming, an example of which I have just presented? Their confidence is based not on evidence, but on faith. At the very least, a little more humility might be appropriate. But as I have noted, the IPCC was never tasked to find alternative explanations for global warming. Its founding purpose was to build the scientific case that humanity is the culprit.

I am not claiming to have disproved manmade global warming with this evidence. No doubt other scientists will try to refute it, and a few will run their own climate models and get different results by making different assumptions. One can get just about anything one wants with computer models if one tries hard enough–although it just so happened that my model explanation worked on the first try.

But the main point here is that the evidence for global warming possibly being mostly caused by natural forcings is sufficiently strong to justify devoting some research effort to investigating the subject. Surely this is better than sweeping the issue under the rug, and then claiming near-certainty that natural sources of global warming do not exist. The IPCC leadership has purposely avoided what is a fundamental tenet of scientific investigation: testing alternative hypotheses.

Why is there so much resistance to the study of potential natural sources of climate change? Judging from the IPCC's history, one can only conclude that it is driven by political motivations and desired policy outcomes. We saw in Chapter 1 that the IPCC has tried to rewrite history, using the hockey stick reconstruction of past temperatures to do away with the Medieval Warm Period and the Little Ice Age, events which recorded history tells us actually happened.

And now the IPCC has systematically downplayed, if not outright ignored, the potential role of nature in recent climate change. Apparently we are supposed to believe the IPCC leaders and their political cheerleaders because they have hundreds of scientists, supercomputers, twenty complex climate models, an Academy Award, and a Nobel Prize.

All I have to offer is scientific evidence. It is unfortunate that any objections to the public proclamations of a "scientific consensus" on global warming have been met with *ad hominem* attacks and ridicule. How dare anyone disagree with the world's leading climate experts?

This would all be of little public interest if it weren't for the policy implications of the IPCC's pronouncements regarding the science. Some politicians are now saying that we need to put control of the global energy supply into the hands of bureaucrats in order to save humanity from destruction.

The IPCC leadership has a history of political activism, demonstrating that something besides a desire for good science is guiding the organization. In the Introduction, I recounted a meeting in the early 1990s with Dr. Robert Watson, the chief environmental scientist for the Clinton-Gore administration. Formerly a stratospheric chemist for NASA, Dr. Watson had helped negotiate the United Nations' 1987 Montreal Protocol to reduce the manufacture of ozone-depleting substances like Freon. Later, he became chairman of the IPCC, from 1997 to 2002. During our visit, Dr. Watson informed me and my associate, Dr. John Christy, that after his success on the stratospheric ozone problem, the next goal was to regulate and reduce humanity's production of carbon dioxide. This occurred before very much climate modeling had been performed. Yet here was a political insider who was instrumental in the newly formed IPCC, telling us that the policy goal regarding CO_2 was already decided! There was no mention of doing any scientific investigation into the possibility that global warming might be more natural than manmade. It had already been concluded that mankind was the cause. The only thing that remained was for an international body to build scientific support for that conclusion.

The IPCC does indeed enlist most of the world's best climate scientists, but these scientists have, in effect, had their hands tied. Their work has been guided by bureaucrats who ultimately decide just who receives government funding and for what kinds of research. The IPCC scientists have been rewarded with continuing government grants and contracts to study manmade climate change – not natural climate change. As for the bureaucrats, they cannot be considered unbiased because their jobs depend on a continuing control over the taxpayers' money, which is their only source of power. The popular opinion that government-funded research is unbiased must be considered quite naïve.

What I have demonstrated with the Pacific Decadal Oscillation is just scratching the surface of naturally induced climate change. What if other modes of natural climate variability—such as El Niño, La Niña, the Atlantic Multidecadal Oscillation (AMO), the North Atlantic Oscillation (NAO), and the Arctic Oscillation (AO)—also contribute to changes in global average cloudiness? It is entirely reasonable to hypothesize that one or more of these does. And if the global cloud cover changes, global temperatures will change as well. Again I emphasize: it would take very small changes in global cloud cover to explain all the temperature variability in the last 2,000 years, shown in Fig. 1. The IPCC's assumption that such small natural variations in global cloudiness do not occur is, in my view, arbitrary and scientifically irresponsible.

I have heard some people say that humans should avoid having any influence on climate whatsoever. Remember, I have not claimed that humans have no influence—only that our influence is small compared with that of nature. Yes, CO_2 is a greenhouse gas, and greenhouse gases on average warm the climate. But the fact that humans might have some small influence on climate should come as no surprise to us. If the existence of trees on the Earth affects climate, then why not the existence of people? Have we really decided to give trees greater rights than humans?

Maybe we can even ponder the unthinkable: what if more carbon dioxide in the atmosphere turns out to be a good thing for life on Earth?

Chapter 7 · CO_2: Dangerous Pollutant or Elixir of Life?

The idea that nature was in a delicate balance before mankind came along is religious, not scientific. Given the necessity of carbon dioxide for life on Earth, we need to consider the possibility that more CO_2 in the atmosphere will be better for life on Earth, not worse.

CARBON DIOXIDE IS necessary for photosynthesis, and thus for life on Earth. For something with such a crucial role, carbon dioxide has a surprisingly small concentration in the atmosphere: only 39 out of every 100,000 molecules of air are CO_2. And humanity's greenhouse gas emissions are so minuscule that it will take five full years of global fossil fuel burning to increase that concentration to 40 out of 100,000 molecules.

These facts give a very different impression from Al Gore's statement that humanity dumps 70 million tons of carbon dioxide into the atmosphere every day as if it were an "open sewer." Gore's disinformation campaign–based on a litany of scientific half-truths, exaggerations, and inaccuracies–helped convince five United States Supreme Court justices in 2007 that the EPA must now consider CO_2 a pollutant. They then directed the EPA to decide whether or not it should be regulated, which depends upon a finding of "endangerment" to human health.

The only reason for believing that carbon dioxide is a "pollutant" that could hurt people is the alleged strong warming effect that extra CO_2 produces. But as we have seen, it's really not the direct warming effect of the carbon dioxide that is of concern, because that effect is too small. The concern over warming is based on the current belief that feedbacks are positive, which means that the climate system is sensitive, which in turn means that nature is going to punish us for putting more CO_2 in the atmosphere by causing massive global warming.

That CO_2 is a greenhouse gas is known beyond any reasonable doubt; and that more of it in the atmosphere should cause some amount of warming is, in my opinion, likely. But the question we should be asking is, *so what?* How significant is our contribution to climate change compared with other, natural sources? In fact, why do we consider all other sources of climate change to be natural, but ours as "unnatural"? All other forms of life no doubt have some small impact on climate, so why not humans? As I have previously asked, if the presence of trees can change the climate system, why not the presence of people? Why do some environmentalists insist on giving greater rights to trees than to humans? The issue is not one of science, but of religion.

We have already seen that the paradigm of climate change represented by the scientific consensus involves an Earth that was unchanging and pristine before humans came along. Many believe that the Earth was in a state of perpetual balance and harmony, and now human pollution has upset that balance. But just how realistic is that romantic view of nature? There are changes occurring in nature all the time, not just because of weather and climate variability, and these changes always involve winners and losers. It has been pointed out that while the concept of a "healthy ecosystem" might have intuitive appeal, it really has no scientific meaning.[1] Most animal life on the Earth is involved in a continuous struggle for existence, and if any balance occurs in an ecosystem it is because a stalemate has been reached where the winners have won, the losers have lost, and now everyone has settled into a protracted war, with a roughly constant flow of casualties on all sides.

You might not be aware that humans are not the only ones changing the amount of carbon dioxide in the atmosphere.

Nature also has a huge influence. Next we'll look at a few of those natural processes.

NATURAL FLOWS OF CARBON DIOXIDE

You have probably been led to believe that mankind now controls the amount of carbon dioxide in the atmosphere. In Fig. 9 of Chapter 3, we saw how the CO_2 content of the atmosphere is slowly rising over time. Since humanity keeps producing CO_2 by burning fossil fuels, it has been assumed that the increase in the atmospheric concentration has been entirely manmade. While this might be true, there are some fairly strong natural influences on how much extra CO_2 shows up in the atmosphere each year.

The estimated annual emissions of carbon dioxide by humanity since 1958 are shown in Fig. 27. If the annual change in the CO_2 concentration in the atmosphere were entirely due to the human emissions, then the measured yearly CO_2 increase at Mauna Loa, Hawaii, would look very similar to the human emissions curve. But as can be seen from Fig. 27, there are huge fluctuations in how much extra CO_2 shows up at Mauna Loa each

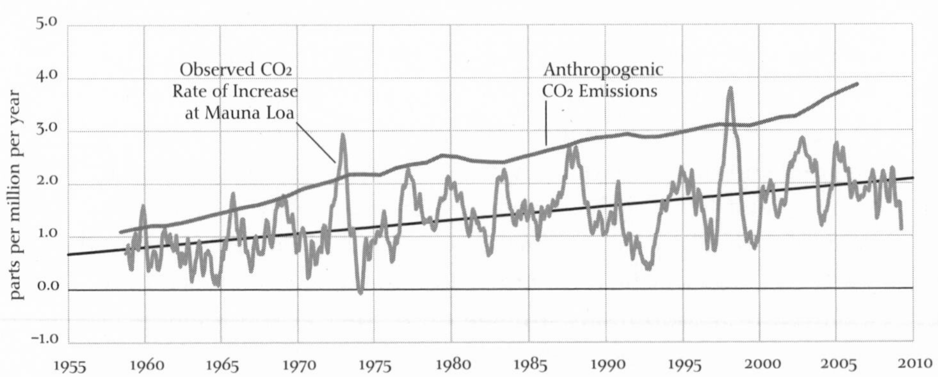

Fig. 27. Estimated yearly anthropogenic CO_2 emissions from fossil fuel use, and the corresponding yearly rate of increase in the atmospheric CO_2 concentration observed at Mauna Loa, Hawaii.

year, indicating that much more than the burning of fossil fuel affects the atmospheric concentration of CO_2.

The reason for these spikes is related to the fact that there are large natural flows of CO_2 in and out of the Earth's surface each year, and these flows are usually somewhat out of balance. The magnitude of the carbon sources and sinks is estimated to be as much as twenty times that of the human emissions. And because these natural flows in and out of the atmosphere are so big, all it takes is a small imbalance between them to cause a yearly change in atmospheric CO_2 that can be as large as the human contribution.

Fig. 27 also shows that nature absorbs much of the carbon dioxide we produce. This is why the yearly rate of increase in the observed CO_2 concentration at Mauna Loa averages only 50 percent of the human rate of emissions. As fast as we can pump more CO_2 into the atmosphere, nature continuously sucks 50 percent of the excess back out again. And since CO_2 is food for the biosphere, this insatiable appetite that nature has for the carbon dioxide we produce begs the question: *Has life on Earth been starved for carbon dioxide?*

For most kinds of plants, more CO_2 in the atmosphere is beneficial, as numerous studies have established.[2] Operators of large greenhouses know that artificially boosting CO_2 concentrations to around 1,000 ppm increases productivity. This is almost three times the current atmospheric concentration of CO_2, currently at about 390 ppm.

The spikes in Fig. 27 are believed to be mostly the result of year-to-year fluctuations in biological activity and changes in ocean temperature. If one compares the ups and downs in Fig. 27 with yearly variations in surface temperature, more CO_2 shows up at Mauna Loa in unusually warm years, and less in cool years.

During unusually warm years, the ocean gives up more CO_2 than it absorbs. For instance, during a strong El Niño, such as that of 1997–1998, more CO_2 is released by the ocean into the atmosphere than is being taken out of the atmosphere by the ocean.

Part of the explanation is the soda "fizz" effect: warm water can hold less carbon dioxide than cold water. But the main reason seems to be that there is less phytoplankton growth in the ocean during El Niño, so less CO_2 is required for photosynthesis.[3] Since phytoplankton are the start of the oceanic food chain, the natural year-to-year fluctuations in atmospheric CO_2 seen in Fig. 27 are evidence that the amount of life on Earth is not static, but subject to the whims of nature.

During the strong 1982–1983 El Niño, changing ocean circulation patterns caused the deaths of literally trillions of marine organisms. This natural climate event had a huge impact on wildlife in the Galapagos Islands, where the populations of some species were decimated, while others experienced explosive growth.[4] Is this the way that a pristine, harmoniously balanced system works?

In contrast, during the cool conditions of La Niña, the combination of cooler waters absorbing more CO_2 and faster plankton growth leads to an anomalously large uptake of CO_2 by the ocean. This is what caused most of the downward spikes seen in Fig. 27. These natural variations in temperature and biological activity have a large impact on how much CO_2 is either released into or absorbed from the atmosphere.

This means that there can be a natural increase in the CO_2 content of the atmosphere as a result of warming. Note that this is opposite to the direction of causation involved in the theory of manmade global warming, wherein the warming is alleged to be the result of the increase in CO_2. So, just as in the case of clouds and temperature, we are once again confronted with the question of cause versus effect.

As far as I have been able to determine, those who do computer modeling of the flows of carbon between ocean and land and atmosphere simply assume there are no natural, long-term sources of CO_2 that might be contributing to the observed increase in the atmosphere. It is assumed to be entirely anthropogenic. In other words, it is assumed that the flows of carbon dioxide in and out of the atmosphere have been in equilibrium for thousands of years, until humans came along and upset the system. Sound familiar?

Am I claiming that the observed increase in atmospheric CO_2 is due to nature rather than mankind? No, I am just asking whether the models might be adjusted in ways that explain part of the increase in CO_2 as natural. Or is that not allowed? I suspect that, just as the IPCC climate modelers cannot bring themselves to admit any substantial role for nature in global warming, the carbon cycle modelers similarly cannot let themselves address the possibility that some of the CO_2 increase over the last fifty years is natural.

There are two main scientific objections to the possibility that increasing CO_2 is partly natural. The first is that the amount of CO_2 that mankind emits each year is more than enough to explain the observed increase. This is shown in Fig. 27 by the fact that the long-term CO_2 growth rate seen at Mauna Loa is only about 50 percent of the rate of human emissions. While this is consistent with all of the increase being anthropogenic, it would also be consistent with a combination of natural and anthropogenic sources if the rate at which some regions are removing CO_2 from the atmosphere is increasing even faster.

After all, researchers have admitted that it is not yet known where all the extra CO_2 is being absorbed, although increased uptake by some combination of the ocean and land-based vegetation seems likely. This has been dubbed the "missing sink" problem.[5] The launch failure of NASA's Orbiting Carbon Observatory (OCO) in February 2009 will hinder the search for the missing sink, although Canada does have an inexpensive, stripped-down version of the OCO satellite now orbiting the Earth. In the coming years, we are likely to learn more about how nature recycles carbon. I predict it will be discovered that nature plays a much larger role in removing excess CO_2 from the atmosphere than scientists currently believe.

Because scientists are not sure where the extra CO_2 is going, why not consider the possibility that some portion of the increase is natural, and that nature is absorbing even more than 50 percent of what we produce each year? How about 60 percent? Or 70 percent? As is the case with climate modeling, I'll bet that computer models of the carbon cycle could be adjusted to make room for a natural source of some of the increasing atmospheric CO_2. But again, that would marginalize the role of mankind, and so I suspect there is considerable scientific peer pressure against performing and publishing any such modeling experiments.

The other supposed proof that increasing CO_2 in the atmosphere is manmade rather than natural is that the tiny fraction of the atmospheric CO_2 that contains the carbon isotope C_{13} rather than the normal C_{12} form of carbon has been slowly decreasing over time. This decrease is claimed to be caused by the burning of fossil fuels, since the CO_2 that results is depleted in the C_{13} isotope in comparison with the CO_2 already in the atmosphere. But it turns out that vegetation and biological activity in the ocean are also associated with lower levels of C_{13}. So a slowly decreasing C_{13} content of the atmosphere is not a unique fingerprint of manmade CO_2; it would also be consistent with a long-term increase in a biological source. Maybe the upwelling of cooler water from the depths of the ocean has slowed in the last fifty years, contributing to a warming of surface waters, a decrease in phytoplankton growth, and therefore extra release of CO_2 by biological activity in the oceans.

Again, I am not claiming that increasing atmospheric CO_2 is mostly natural. I am merely asking, are there other potential explanations for the observed CO_2 changes where mankind is not the only culprit? The modelers like to claim that their computer model explanations are "consistent with" humans causing all of the CO_2 increase (or all of the warming). What they don't tell you is whether there are other model explanations that include a role for nature which are also consistent with the observations. If such model experiments are being performed, they are not being released for public consumption.

This is important because if some portion of the recent increase in atmospheric CO_2 is natural, it means that the extra CO_2 we are putting into the atmosphere is being absorbed faster than we thought. This in turn would suggest that when we finally do find large-scale replacements for fossil fuels, nature will be gobbling up the extra CO_2 out of the atmosphere faster than is currently projected.

Ocean Acidification

Even if an increase in atmospheric carbon dioxide does not cause global warming, there remains another worry: Increasing atmospheric CO_2 is believed to be causing "ocean acidification" as some portion of the extra CO_2 is absorbed by the ocean. The term "ocean acidification" is a misnomer since the pH of seawater is solidly in the alkaline range, with an average estimated to be about 8.10 today. It is believed that this is down from a pH of about 8.18 in preindustrial times, but no one really knows for sure. There is a wide range of ocean pH values across the world's oceans, and until recently actual measurements were quite sparse.

The main concern over having more atmospheric carbon dioxide available to the ocean is its effect on sea life. This is still a poorly understood subject, and much research is being performed to get a better idea of the future consequences as CO_2 concentrations continue their slow increase. But given the fact that science continually underestimates the ability of nature to adapt, I would wager that science will eventually establish that more atmospheric CO_2 is actually good for sea life, just as it is good for vegetation on land. In fact, recently published evidence suggests that this is already happening, with more vigorous growth of plankton.[6]

Also, there is new evidence that the excess carbon is beginning to show up on the ocean bottom in the form of a slight increase in the rate at which dead organic matter precipitates out of the ocean after making its way through the oceanic food chain. Indeed, this could be the ultimate fate of most of the extra CO_2 that we generate. If all the CO_2 that humanity produces by burning fossil fuel each year ended up being deposited as carbon on the sea floor, it would add a layer 0.04 millimeters thick (about 0.002 inches). That is the thickness of a single human hair. Of course, this would be in addition to all the dead organic matter that ends up being deposited on the ocean floor each year anyway.

The effect of more atmospheric CO_2 on the oceans is still quite uncertain, but since the oceans seem to have survived geologic periods with much higher atmospheric concentrations of CO_2,[7] I suspect that ocean acidification, like anthropogenic global warming, will also end up being a false alarm.

I would wager that there will be many more winners than losers in the oceans, but it will probably take a long time before science provides the evidence to support this proposition. In the meantime, you can be sure that most of what is reported through the news media on this subject will be dominated by the views of alarmists, just as is the case with global warming. Any scientific research indicating that the oceans will benefit from more atmospheric CO_2 will be either marginalized or outright ignored by the news media.

SHOULD WE BE CUTTING BACK ON OUR CO_2 EMISSIONS?

Let's assume for the sake of argument that all of the increase in atmospheric carbon dioxide is caused by humans, and maybe even that atmospheric CO_2 really is higher now than it has been for hundreds of thousands of years. My question is, again, "so what?"

Just because nature has found a particular state of balance does not mean it is the preferred state. For instance, what if life on Earth really has been starved for atmospheric CO_2? After all, as shown in Fig. 27, no matter how much more we put into the atmosphere each year, life on Earth gobbles up about 50 percent of the excess. Maybe we are doing nature a favor by adding more of this essential nutrient to the atmosphere.

Here's a little thought experiment: Imagine that the amount of sunlight reaching the Earth was much less than it is today, and that global temperatures were much lower as a result. Some forms of life could not survive the colder conditions, and the total number of plant and animal species would probably be reduced. But at some point, a natural state of equilibrium would still have become established, and that equilibrium state might last for hundreds of thousands of years.

Then imagine that humans come along and declare nature to be in a harmonious state of balance, not realizing that most forms of life would actually have preferred the Earth to be a little warmer, thank you. How is this any different from the situation we have today?

Just because a natural state of balance has existed for so long, does that mean it is in any way a preferred state? It just represents the state at which the resulting populations of plants and animals have finally settled into a state of mutually agreeable quasi equilibrium. It might not be the balance that most of the different species of plants and animals would have chosen individually, but it is the one they are forced to live with anyway.

Or, imagine that human activities were destroying atmospheric CO_2 rather than creating more of it. There would be howls of protest from environmental groups that we were depriving life on Earth of the very food it needed to survive. Climate modelers would warn that we were driving the global climate system toward another ice age.

The idea that any state of balance in nature is a preferred one is philosophical or religious, not scientific. I have even seen scientific journal articles refer to the "delicate balance" of nature. Where did we get the idea that any state of equilibrium is "delicate"? In my opinion, such views amount to nature worship. Don't get me wrong: nature worship is fine from a freedom-of-religion standpoint. It is protected by the free exercise clause of the First Amendment to the U.S. Constitution. But when it leads to legislation or regulations to restrict CO_2 emissions, it verges on a violation of the establishment clause of the First Amendment: "Congress shall make no law respecting an establishment of religion."

The public debate over carbon dioxide needs to be reframed. Instead of asking *By how much should we cut back our CO_2 emissions?*

we should ask *Is there any compelling reason to reduce CO_2 emissions at all?*

The argument that more CO_2 might have negative impacts on some portions of the environment is irrelevant. When natural climate change happens, there are always winners and losers. Why should the human impact on nature be expected to be any different? If humans cause some small amount of warming, why should we necessarily be opposed to it? Especially since, at least for the time being, there is little that can be done about it?

The debate over global warming would not be nearly as contentious if it involved some chemical whose emissions we could easily reduce. For instance, we greatly lowered our emissions of sulfur dioxide by putting scrubbers on the smokestacks of coal plants. But as yet there are no scrubbers that can take out carbon dioxide–at least not on the scale of what would be required for mankind's rate of energy use.

Global warming alarmists frequently point to the stratospheric ozone issue as proof that we can use government regulation to reduce CO_2 emissions substantially. Manmade chlorofluorocarbons (CFCs) were invented because they have excellent refrigerant properties. But as the evidence mounted that CFCs might be destroying some of the stratospheric ozone that shields life on Earth from damaging ultraviolet radiation, the 1987 Montreal Protocol was signed to reduce the manufacture of CFCs and other ozone-depleting substances.

It is argued that since we fixed the CFC problem, we can therefore fix the CO_2 problem. But there is no basis for comparing CFCs and CO_2 on an equal footing. Whereas CFCs were manmade chemicals that could be replaced with less harmful alternatives, CO_2 is a natural and necessary component of life on Earth. It is the unavoidable byproduct of all kinds of natural processes in addition to our use of energy. There are no alternatives to carbon dioxide, and fortunately life on Earth loves it.

If it were relatively easy to reduce our CO_2 emissions, then it would make sense to work toward that goal. But humanity's demand for energy is so large – and increasing – that the only way to meet most of that demand is with fossil fuels. Short of shutting down the world economy, there is very little that humans can do to reduce CO_2 emissions substantially – at least not until a new energy technology is developed. Wind and solar power can help a little, as can more nuclear power, but fossil fuels will dominate our energy mix for decades to come.

But politicians are now creating a public perception that carbon-based fuels can easily be replaced with renewable alternatives on a substantial scale. They claim that the energy industry is just being lazy by not embracing renewable energy technologies fast enough. Not only can carbon-based fuels be replaced, it is claimed, but we have a moral and patriotic duty to reduce our use of fossil fuels. All humanity will benefit, and the economy will be invigorated by the creation of many new green industries and jobs.

If that is true, it will happen anyway. There is no need to legislate something that is economically beneficial in a free-market economy. And since the demand for energy by humanity is so great, the free market will also ensure that cost-competitive alternatives to fossil fuels are developed. If nothing else, the increasing cost of finding and extracting enough fuel from the ground to supply the growing global demand for energy will force the development of new energy technologies.

Big Business

Companies like British Petroleum (BP) do not help matters when they run TV commercials that make it appear they are "going green." In late 2008, BP cancelled plans for renewable energy projects in Britain due to the high cost.[8] They will continue investing in renewables in the United States, but only because the U.S. government is subsidizing them. Of course, subsidies are merely a way to force taxpayers to prop up industries that cannot compete economically in the free market. BP also decided to expand into

Canadian tar sands, a move strongly criticized by Greenpeace because a lot of extra energy is required to turn tar sands into fuel.[9] But for some reason this environmentally stressing source of petroleum never made it into BP's "green" commercials.

While pandering to the public's environmental anxieties might be an effective marketing strategy, I consider it detrimental in the long term. When Big Business poses as being on the "CO_2 is evil" bandwagon, the public perceives it as an acknowledgment that CO_2 is a real problem and that something must be done about it. In reality, this is just clever marketing to make industries and companies appear greener than they really are. I'm not against big business, but I am against playing on myths and public ignorance, thereby giving support to legislation or regulations that will end up hurting people. If the CEOs of all petroleum companies, coal companies, electric utilities, and heavy industry eventually come out in favor of either carbon cap-and-trade legislation or a carbon tax, it will be a result of political and strategic calculations as they try to position their companies for what many see as inevitable restrictions on fossil fuel use. Those decisions will not be based on what would be most beneficial to humanity.

I am currently very supportive of fossil fuels because I know that we really have no adequate alternatives. I recognize the importance of abundant and affordable energy to help eliminate the greatest scourge that humanity faces: *poverty*. Forcing expensive alternative forms of energy on people in the futile attempt to fight global warming is nothing less than a war on the poor by those who are wealthy enough to pay higher prices for energy. That is why Roy Innis, the national chairman of the Congress of Racial Equality, has called this "the new civil rights battle."[10]

I'm sure that we will eventually develop cost-effective and widely deployable alternatives to fossil fuels. But these replacements cannot be legislated into existence. Just because we put a man on the moon doesn't mean we can now construct a time machine, or a transporter that can move you to the other side of the world in the blink of an eye. Technology has done wonderful things for humanity, but that does not mean it can do anything we want.

Research into energy alternatives is already being carried out by our government at considerable expense to the taxpayer, and by private companies at considerable expense to the consumer. This is how we solve technological problems as they arise, and since everyone needs energy, you can bet that every potential solution is being investigated. Yet the political pundits give the public the impression that government and the energy companies are just sitting on their hands.

One aspect of the global warming issue that I find particularly disturbing is the reception I get from the environmentally concerned when I describe the evidence for global warming being natural rather than manmade, or my claim that more carbon dioxide in the atmosphere might be a good thing. The reaction ranges from frowns to scorn to insults. Pardon my ignorance, but shouldn't dodging the global warming bullet be considered a good thing for humanity? At least you would *think* it's good news. But this is not the way many people greet the evidence provided by "skeptics." Rather than hear comments like, "Wow! That would be great news," scientists like myself hear remarks more along the lines of, "You're obviously a shill for Big Oil," or, "You're just like the scientists who were paid off by tobacco companies to say that cigarette smoking wasn't dangerous." Oh, really? Where is the evidence that anyone has ever been hurt by manmade climate change?

There is no reasoning with such people. They are determined to be miserable, and no one is going to talk them out of it. They seem to be *hoping* that global warming is a major threat to humanity. And for some reason, they also seem to be the ones who have benefited the most from the industrial age, modern conveniences and medical care, and they have the leisure to worry about our energy use hurting the planet. They've got theirs, and now they want to deprive poor people in other countries of the same opportunity to lift themselves out of poverty.

Yes, I look forward to the day when we can begin a public

dialog on the harmful consequences of making energy more expensive, and maybe even discuss what I regard as a very real possibility: *More atmospheric carbon dioxide might be good for life on Earth.* We have enough real problems to address in this world without making up imaginary ones.

Chapter 8 · Out on a Limb: Predictions for the Future

I HAVE USUALLY resisted making climate predictions. But since everyone else is making predictions, I suppose I can climb out on a limb and give it a try, too. What follows are some potential outcomes regarding future global temperatures, global warming science, and energy policy. They range from naïvely fantastical to depressingly cynical. I suspect that what eventually happens will be somewhere in between.

FUTURE TEMPERATURES

First let's look at the future of global temperatures. One rather necessary inference from the new science I have presented is that global warming will either stop in the coming years – if it has not already done so – or proceed at a much slower rate than is being projected by the IPCC. At this writing there has been no warming for eight years (since 2001), and there is no sign as yet that warming will resume anytime soon.

Much of this book has focused on the possibility that the Pacific Decadal Oscillation has caused most of our global average temperature variability in the last century. Since the PDO changes phase approximately every thirty years, it would be about time for a new, negative phase to take over. A few experts believe this has already happened. The main proponent of this view is Don

Easterbrook, a geologist at Western Washington University who, like me, believes that the PDO has been the main driving force of global temperatures for the last 100 years or more.

If the PDO continues in its negative (cooling) phase, then some cooling might be expected for the next twenty or thirty years. But since the extra carbon dioxide that humanity produces probably has some warming influence, the PDO-induced cooling would be partly cancelled out by anthropogenic warming, leading to a prolonged period of little temperature change. The evidence I have presented for low climate sensitivity (negative feedbacks) would indicate that the long-term warming from the extra CO_2 will be small in any case. While the IPCC is 90 percent sure that global warming from a doubling of atmospheric CO_2 will not be less than 1.5 deg. C, at this point I would put that probability closer to 50-50.

This is all speculative, of course. My main purpose in this book is not to claim that the PDO necessarily constitutes the largest single mechanism of climate change – although that is a possibility. Instead, my aim is to demonstrate that the "scientific consensus" that global warming is caused by humans is little more than a statement of faith by the IPCC. There is evidence of natural climate change all around us if scientists would just take off their blinders.

Now let's turn to what might happen in the scientific and political realms. I'll discuss the fantasies first.

The IPCC Has an Epiphany

One remote possibility is that the IPCC leadership will ask its scientists to look into the possibility that Spencer or a number of other skeptical scientists have a valid argument. The IPCC scientists, in turn, finally understand what we've been talking about all these years. The moment of epiphany arrives with one of the leading IPCC scientists declaring something like, "By Jove, I think he has a point, Professor!"

All of the scientists, politicians, bureaucrats, and governmental representatives connected with the IPCC then heave a collective sigh of relief as they realize that humanity is saved from the ravages of anthropogenic global warming.

Al Gore apologizes for the whole misunderstanding, and returns his Academy Award for best science fiction movie, as well as his Nobel Peace Prize.

The IPCC leadership, in true scientist form, point out that they never did claim there was a 100 percent chance that global warming would be serious – only 90 percent. Therefore, technically speaking, the outcome was consistent with their predictions.

This is all just a fantasy, of course, but it was fun while it lasted.

An IPCC Scientist Has a "Blue Dress" Moment

A somewhat more likely outcome is the scenario where a scientist central to the IPCC effort either flips to our side, or accidentally spills the beans regarding what the IPCC has been hiding on the subject of natural climate change. Chris Horner has called this potential event the "blue dress" moment, with the IPCC's dirty laundry finally coming to light, possibly through the altruism – or mistake – of a single person.

Maybe an IPCC scientist, through the inherent curiosity that scientists used to be known for, starts looking into natural climate variability rather than sweeping it under the rug. But the scientist's boss finds out about it and tells the researcher in no uncertain terms to stop rocking the boat.

Since scientists don't like being told what to do, the rogue scientist continues to investigate, researching the latest satellite observations of the Earth with a simple climate model on his home computer. (This is beginning to sound like me. I wonder who will play me in the movie version. Maybe Steve Martin – we have similar senses of humor.)

Finally, in a fit of mental clarity and scientific objectivity, with a newly informed understanding of how the world really works after reading Thomas Sowell's *Basic Economics*, the stubborn scientist writes up his research and submits the results for publication.

This is the point where my second fantasy gets derailed. Still under the illusion that journalists are objective in such matters, the scientist tells a science reporter at one of the nation's major newspapers about his new results. An email is then quickly dispatched from the newspaper to Earth First!

The story ends with a small obituary in the local newspaper for the prominent, Nobel Peace Prize–sharing scientist who died in a freak accident while filling his hybrid with gas. The Pew Charitable Trusts honors the scientist with a special scholarship in his name that will provide financial assistance to environmentally minded students going to journalism school.

Since I first wrote the previous few paragraphs, the unauthorized "Climategate" release of over a thousand emails in November 2009 has revealed that the core group of IPCC scientists responsible for the surface temperature record engaged in discussions to delete, hide, and manipulate temperature data, as well as to interfere with the peer review process to favor the IPCC's objectives. Only time will tell if any lasting damage to the IPCC's reputation has resulted from Climategate.

INDEPENDENT REVIEW OF THE IPCC AND CLIMATE RESEARCH FUNDING

Still higher on the probability scale for the future is an independent review of the IPCC process. Given the huge cost of implementing the policies that the IPCC's conclusions would demand, such a "red team" approach might be wise. I think it is more than a little risky to trust the United Nations to be an objective arbiter of an issue with such huge political and financial ramifications for the countries of the world.

As I noted earlier, the IPCC was not chartered to investigate possible natural sources of climate change, but rather to build the scientific case for mankind being the cause of global warming. The IPCC scientists will claim they can't think of anything else that might be the cause, but no serious effort was ever expended to look for explanations in nature. I believe they are therefore in no position to cast judgment on whether global warming is mostly manmade or mostly natural.

Of course, any independent review of the IPCC's activities will be heavily criticized and lobbied against. How dare politicians question the world's leading scientists and listen to the lunatic ramblings of pseudo-scientific flat-Earthers? Because of this pushback, there would have to be strong public support for a review of the IPCC to take place. Politicians are reluctant to go up against scientific organizations, but they will if their constituents demand it. Given the huge cost to the consumer of any government-mandated regulation of CO_2 emissions, I hope that this kind of public interest will be cultivated at the grassroots level.

In a review of its activities, the IPCC leadership would likely defend its position by pointing out that little or no published evidence exists for a natural source of global warming. But this is primarily because the research community has not been funded to look for natural sources of climate change. Maybe we should find out what fraction of the total amount of taxpayer money spent on climate research in recent years has gone into investigating possible natural sources of climate change. I would guess that this fraction is very close to zero. In reality, the IPCC's position is not based on the evidence, but on a lack of evidence. For too long the panel has succeeded in deflecting criticism while hiding behind a veil of professed scientific objectivity.

Unfortunately, the climate research community is relatively small, so it would be difficult to put together a truly independent review of the IPCC. Yet there are some scientists from other disciplines who have a sufficient grasp of the physical concepts involved in climate change to preside over such a review. I regularly hear from physicists, chemists, and engineers who are distrustful of the IPCC's claims. Even though they are not climate experts, they still know enough about how the natural world behaves to come up to speed on the basic issues and then intelligently critique those claims. I am hopeful that this book will encourage an independent review with the participation of professionals from related fields of research.

Public Debates on Global Warming

I would like to see one or more well-publicized debates between representatives from both the global warming alarmist and the skeptic camps. Al Gore purposely avoids any challenges to debate the issue. This is probably a good strategic move for him because he would run the very real risk of losing.

Debating would also serve to legitimize the skeptics' side of the issue, which is clearly not in the best interests of the alarmists. While refusing to debate, Mr. Gore can say things like, "What's the point of debating whether the Earth is flat or round? We already know the answer to that one." Even in his Senate testimony of March 21, 2007, he refused to answer any scientific objections put forth by Senator James Inhofe, other than to state his desire to sit down with the senator in private so that reason might prevail.

If a public debate were held, how would a winner be determined? I like the idea of surveying the opinions of the audience before and after the debate, in order to get some idea of the persuasiveness of the debaters' arguments. This was actually done in an Oxford-style global warming debate in New York City on March 14, 2007. Those supporting the motion "global warming is not a crisis" were the late novelist Michael Crichton; Richard Lindzen, a professor of meteorology and climate researcher at MIT; and Phillip Stott, an emeritus professor of biogeography at the University of London. Those opposing the motion were Brenda Ekwurzel, climate scientist with the Union of Concerned Scientists; Gavin Schmidt, a climate modeler at the NASA Goddard Institute for Space Studies; and Richard C. J. Somerville, professor at Scripps Institution of Oceanography.

Prior to the debate, a survey of the audience revealed that about 30 percent of the audience agreed with the motion that "global warming is not a crisis," while 57 percent were against the motion, and the remainder were undecided. But after the debate, the number who agreed with the "skeptics" grew to 46 percent, while those on the other side shrank to 42 percent. This result shows that when people are exposed to the scientific evidence that has been hidden from them, they begin to question the consensus.

Such a debate could even be televised and judged by the public over the internet. For instance, only those who participated in a simple online survey before the debate would be allowed to take part in a follow-up survey after the debate, using the common online polling method of keeping track of computer IP addresses. It has also been suggested to me that the public visibility of a debate could be enhanced by the participation of betting in the United Kingdom on the outcome.

Unfortunately, the voting for such an event can be rigged. Both sides can pretend to have been converted from their opponents' side, thus skewing the numbers. Also, in my experience, organizers of anything that looks like a debate have been unable to enlist representatives from the IPCC to participate. It is my understanding that James Hansen won't participate because he considers such efforts to be a waste of time that would be better spent trying to convince everybody of the "climate crisis." On several occasions, the organizers of my own public lectures have had great difficulty finding anyone to present the other side of the issue. They usually give up after two or three rejections. I have even been known to help out my opponents because they did such an awful job of presenting the IPCC's side.

If the skeptics' arguments are so ridiculous, then why would the alarmists not want to confront them publicly and expose their folly? It is unfortunate that the Keepers of All Climate Knowledge instead use *ad hominem* insults and a variety of propaganda techniques to stifle debate on the subject. I find such tactics quite offensive, and I hope you do, too. Let's see if they can enter the arena and defend their claims on the science alone.

* * *

EPA Regulations and Congressional Legislation

For every year that goes by without a resumption of global warming, it will become increasingly difficult for the IPCC and their representatives to scare the rest of us into taking action to combat global warming. As I showed in Chapter 2, their claims have already become so shrill that they are almost comical.

Oh, the alarmists might spring back into action whenever there is a tornado outbreak, or a Category 5 hurricane, or a heat wave, as if such events never happened before we started driving SUVs. But their influence will continue to diminish as the public realizes that repeated forecasts of future warming keep getting postponed. Indeed, if we have entered into a new, negative phase of the Pacific Decadal Oscillation, we might see thirty or more years of no warming – or maybe even cooling.

Some of us suspect that one of the reasons why there has been such a strong push to pass carbon legislation in Congress is the realization that a continued lack of warming might cost the alarmists their only chance for legislative success. As long as the global economy remains weak, it will be difficult to legislate reductions in greenhouse gas emissions. But the Democrat-controlled Congress still could pass legislation in 2010 to cap carbon dioxide emissions.

Legislation is not the only way to force reductions in CO_2 emissions. Since the Supreme Court has told the EPA that it must consider carbon dioxide a "pollutant" under the Clean Air Act, another path is for the EPA to regulate sources of CO_2. Given President Obama's expressed interest in phasing out coal and forcing increased reliance on green energy sources, new EPA rules to regulate CO_2 emissions may well be implemented even if global warming does not resume. Barack Obama even predicted that the new emphasis on clean energy will make it very difficult for electric utilities to build any more coal-fired power plants, because any attempt to do so under new federal regulations would drive them into bankruptcy.

A number of states – and especially governors of those states –

have decided not to wait for action at the federal level. They have instead taken it upon themselves to lead the way in reducing CO_2 emissions. California, always a pacesetter in progressive issues, has mandated lower CO_2 emissions from cars than federal standards dictate. Because of California's large population, this might end up becoming a *de facto* federal standard, which could then cause additional turmoil in the American car industry when it is already having great difficulty competing in the global market.

In 2007, through the efforts of its governor Kathleen Sebelius, Kansas became the first state to block the construction of a coal-fired power plant over fears of global warming. Again in 2009 Sebelius vetoed legislation that would have allowed their construction. But business owners in Kansas and other states are genuinely concerned that the financial burden associated with the regulation of CO_2 emissions in their state might drive businesses to other states or even cause them to fail altogether.

It might seem that such concerns would be alleviated if the United States implemented regulations at the federal level rather than the state level, but in that case many businesses would relocate to other countries. Since our environmental regulations are more stringent than those in most other countries, chances are that these industries would then be able to pollute even more than if they had stayed in the United States. More (alleged) harm would then be done to the environment than if those federal regulations had not been imposed!

So, the way around *that* problem is to have all the nations of the world agree on regulations to reduce CO_2 emissions. But global regulation of CO_2 emissions seems far off at this writing, mainly because India and China are rapidly growing their economies and refuse to participate. As of 2008, China has overtaken the United States as the global leader in CO_2 emissions. China argues that because the United States is the main source of the excess CO_2 already in the atmosphere, it should bear most of the burden in emissions reduction. This argument is not without some merit. But again note that everyone has just *assumed* that our CO_2 emissions are the cause of global warming.

If the economy does suffer from new legislation or EPA regulations on CO_2 emissions, there will no doubt be a concerted effort to blame the downturn on something else. Your attention will be distracted by claims that Big Business, not the government, has caused the problem by not embracing alternative forms of energy quickly enough.

But you cannot legislate new energy sources into existence. The technologies required to reduce CO_2 emissions substantially, say by 50 percent or more, do not exist yet. Solar energy would be great if not for the fact that it disappears at night, it is practical only in mostly cloud-free areas of the country, and when deployed on a large scale it has its own negative impact on the environment. Putting solar collectors in geostationary orbit around the Earth would provide a continuous source of energy, day and night, but it would be insanely expensive per kilowatt-hour of electricity. Wind energy is great, too, as long as the wind is blowing. But like solar energy, it has to be backed up with fossil fuel.

So, while regulations and legislation might be intended to make things better for people, they are likely to make things worse.

One of the practical problems faced with cap-and-trade legislation is the decision of how many carbon credits should be given out to businesses. Since companies will have to be allowed to emit a certain amount of CO_2, they must be issued permits by the government. Who will decide how many permits will be given to each industry or to each company? What kind of creative carbon-accounting techniques will be used by companies to inflate their claimed need to use energy? What kind of special allowances will politicians give to their favored industries?

The whole cap-and-trade process is a breeding ground for new sources of cheating and corruption that haven't even been invented yet. It sounds like a bureaucrat's dream. It is advertised as a "market-based" approach to reducing greenhouse gas emissions, but it is really a tax in disguise, as consumers end up paying the extra costs. Don't be fooled by politicians who make it sound like a free-market approach to reducing CO_2 emissions.

In fact, cap-and-trade has a history of not even helping to reduce CO_2 emissions. It amazes me that our public discourse on CO_2 legislation and regulation has so far ignored the failure of Europe's emissions trading scheme. Europe and the UK have discovered just how damaging a cap-and-trade approach is to their economies. Inequities between countries result in private industries moving their operations to other countries where they can be more profitable. These problems all stem from the fundamental mismatch between market forces, which always act to maximize productivity and wealth generation, and carbon control legislation, which inevitably imposes limits and extra costs on productivity.

The Public Gets Informed

My favorite scenario involves the public becoming much better informed on how they have been misled by the IPCC, by the media, and by politicians. We need a free flow of information and ideas in order to prosper, so in today's high-tech world the public must become informed on issues they have never had to worry about before. Global warming is one of those issues.

In this book I have tried to bring the science of global warming down from its ivory tower to a level where the public can understand the most important uncertainties. Specifically, I have addressed what I consider to be the core questions that the climate modelers have avoided for too long. While the evidence I have presented may be appreciated most by physicists and engineers, I have included it in the hope of forcing these scientific issues out into the open.

I have attempted to simplify the issues enough for the public to understand because I see a disturbing trend in science: using complex computer models as the ultimate source of scientific evidence. This is dangerous because it is all too easy to manipulate models so they support preconceived notions – or desired policy outcomes.

Even if the U.S. Senate passes carbon cap-and-trade legislation, or the EPA implements regulations to limit CO_2 production, the public will eventually realize that these misguided efforts cause immense economic pain for no gain. Maybe then we can revisit the issue of global warming and ask, How good is the science behind the theory that humans are causing it?

I hope that the information in this book will empower people to start asking some hard questions of their elected representatives. Taking my message directly to the people seemed like a necessary step, because I have found that even publishing peer-reviewed research is no guarantee that anyone will take notice. The environmental lobby has been so successful at shaping public opinion on global warming that even the heads of some corporations have given in to the pressure.

The whole subject of climate change has been made so complex that the public and the politicians can do little but compare the numbers of news reports supporting anthropogenic global warming as a serious threat versus those suggesting that it's a false alarm. If that is the measure of scientific truth, then my side loses. But as I have pointed out earlier, one scientific study by itself could demolish the theory of manmade global warming. That's the way science works.

This book is my attempt to cut through all the peripheral issues and focus on the central uncertainty in the global warming debate. I have tried to boil the matter down to the glue that holds the theory of anthropogenic global warming together: feedbacks. Without a sensitive climate system dominated by positive feedbacks, the case for manmade global warming evaporates. And as I have demonstrated, there is ample evidence that the climate system is dominated by negative feedbacks. In other words, our climate does not particularly care how big your carbon footprint is.

I fully expect that the more successful I am at influencing public opinion and our elected representatives in government, the more I will be attacked. So far, the list of claims against me has been fairly short. One is that I have been bought off by Big Oil, which is flatly untrue. My research has always been 100 percent supported by the U.S. government. I have never been asked by a petroleum company to do anything for them, let alone gotten paid for it. While I have given talks at electric utility conferences, I have also given talks at environmental conferences. I'm an equal-opportunity speaker.

In contrast, the environmentalist lobbyists have been heavily funded by people who support specific political goals and policy outcomes. These leanings are almost always against free markets and against big business. While petroleum companies continue to provide goods that are demanded by most of humanity, many environmental interests would fade away if the threat of global warming were to disappear.

Petroleum companies will survive with or without environmental concerns, since they will continue to provide commodities that everyone needs. In contrast, the existence of environmental advocacy groups depends on a constant stream of environmental fears.

I'm not against trying to minimize the pollution we produce. I am against using the courts and Congress to sacrifice human lives at the altar of religious environmentalism. And the better informed the public becomes – even in the face of a vast disinformation campaign – the less damage the global warming alarmists and their sympathizers in government will inflict on the economy and on society.

Summary & Conclusions

IN SCIENCE, it takes only one finding to overturn decades of mainstream belief. Scientific knowledge is not a matter of consensus, as if scientific truth were something to be voted on. It is either true or not true. I have described new and important scientific evidence – some published, some unpublished at this writing – that supports two major conclusions that could end up dismantling the theory of anthropogenic global warming.

The first conclusion is that recent satellite measurements of the Earth reveal the climate system to be relatively insensitive to warming influences, such as humanity's greenhouse gas emissions. This insensitivity is the result of more clouds forming in response to warming, thereby reflecting more sunlight back to outer space and reducing that warming. This process, known as negative feedback, is analogous to opening your car window or putting a sun shade over the windshield as the sun begins to heat the car's interior. An insensitive climate system does not particularly care how much we drive SUVS or how much coal we burn for electricity.

This evidence directly contradicts the net positive feedback exhibited in the computerized climate models tracked by the IPCC. It is well known that positive feedback in these models is what causes them to produce so much warming in response to humanity's greenhouse gas emissions. Without the high climate sensitivity of the models, anthropogenic global warming becomes little more than a minor academic curiosity.

The strong negative feedback in the real climate system has not been noticed by previous researchers examining satellite data because they have not been careful about inferring causation. As is the case in all realms of scientific research, making the measurements is much easier than figuring out what those measurements mean in terms of cause and effect. Climate researchers have neglected to account for clouds causing temperature change (forcing) when they tried to determine how temperature caused clouds to change (feedback). They mixed up cause and effect when analyzing year-to-year variability in clouds and temperature. You might say they were fooled by Mother Nature. Clouds causing temperature to change created the illusion of a sensitive climate system.

In order to help you understand this problem, I have used the example that I was given when I asked the experts how they knew that feedbacks in the climate system were positive. It was explained to me that when there is an unusually warm year, researchers have found that there is typically less cloud cover. The researchers assumed that the warming caused the decrease in cloud cover. This would be positive feedback because fewer clouds would let in more sunlight and thereby amplify the warming.

But I always wondered: How did they know that it was the warming causing fewer clouds, rather than fewer clouds causing the warming? As we have seen, they didn't know. And when the larger, contaminating effect of clouds causing temperature change is taken into account, the true signal of negative feedback emerges from the data. I have demonstrated this with a simple climate model by showing that the two directions of causation – forcing and feedback (or cause and effect) – have distinctly different signatures both in the satellite data and in a simple model of the climate system. These distinct signatures even show up in the climate models tracked by the IPCC.

Probably as a result of the confusion between cause and effect, climate models have been built to be too sensitive, with clouds erroneously amplifying rather than reducing warming in response to increasing atmospheric carbon dioxide concentrations. The models then predict far too much warming when the small warming influence of more manmade greenhouse gases is increased over time in the models. This ultimately results in predictions of serious to catastrophic levels of warming for the future, which you then hear about through the news media. While different models predict various levels of warming, all of them exhibit positive feedbacks. The mix-up between cause and effect also explains why feedbacks previously diagnosed from satellite observations of the Earth by other researchers have been so variable. There have been differing levels of contamination of the feedback signal by forcing, depending on what year the satellites were observing the Earth.

The second major conclusion of this book is closely connected to the first. If the carbon dioxide we produce is not nearly enough to cause significant warming in a climate system dominated by negative feedback, then what caused the warming we have experienced over the last fifty years or more? New satellite measurements indicate that most of the global average temperature variability we have experienced in the last 100 years could have been caused by a natural fluctuation in cloud cover resulting from the Pacific Decadal Oscillation (PDO). Nine years of our best NASA satellite data, combined with a simple climate model, reveal that the PDO causes cloud changes that might be sufficient to explain most of the major variations in global average temperature since 1900, including 75 percent of the warming trend.

Those natural variations in clouds may be regarded as chaos in the climate system – direct evidence that the Earth is capable of causing its own climate change. Contrary to the claims of the IPCC, global warming or cooling does not require an external forcing mechanism such as more greenhouse gases, or a change in the sun, or a major volcanic eruption. It is simply what the climate system does. The climate system itself can cause its own climate change, supporting the widespread public opinion that global warming might simply be part of a natural cycle. I am not the first to suspect that the PDO might be causing climate change. I just took the issue beyond suspicion, with a quantitative explanation based on both satellite observations and some analysis with a simple climate model.

While some might claim that the timing of the PDO and associated changes in cloudiness in recent years is just a coincidence, I can make the same claim for the supposed anthropogenic explanation of global warming: Just because warming in the twentieth century happened during a period of increasing CO_2 in the atmosphere doesn't necessarily mean that the increasing CO_2 caused the warming. In fact, the PDO explanation for warming actually has a couple of advantages over the CO_2 explanation.

The first advantage is the fact that variations in cloud cover associated with the PDO actually "predict" the temperature changes that come later. It just so happens that the three PDO changes that occurred in the twentieth century were exactly what would be needed to explain most of the temperature changes that followed: warming until the 1940s, then slight cooling until the 1970s, and then resumed warming through the 1990s. This then answers a question I am sometimes asked: How do I know that the PDO-induced cloud changes caused the temperature changes, and not the other way around? It's because the temperature response comes after the forcing, not before. This PDO source of natural climate change can also explain 75 percent of the warming trend during the twentieth century. Addition of CO_2 and other anthropogenic and natural forcings can explain the other 25 percent.

This investigation took me only a few days with a desktop computer. In contrast, researchers have been tinkering for many years with various estimates of manmade aerosol (particulate) pollution in their attempts to explain why global warming stopped between 1940 and the late 1970s, even though this was a period of rapid increase in our greenhouse gas emissions. So, while the PDO explanation for temperature variations during the twentieth century fits like a hand in a glove, the IPCC's explanation based on aerosol and greenhouse gas pollution had to be wedged in with a crowbar.

Another advantage of the natural explanation for global

warming is that the mechanism – an energy imbalance of the Earth caused by natural cloud variations – was actually *observed* by satellite. In contrast, the cooling effects of aerosol pollution and the warming effects of greenhouse gas emissions have remained too small to be measured. They have to be calculated theoretically before being input into climate models.

The process through which I made these findings started with a hypothesis: Natural changes in global cloud cover associated with the PDO might explain global average temperature variability during the last 100 years. Then I used a simple climate model to see if my hypothesis was even possible from a theoretical standpoint, and if so, how much of a change in clouds would be necessary. Last, I turned to satellite observations of the Earth to see if there really was a change in cloudiness that was consistent with the model prediction. The satellite data revealed that there was.

And if one person with an idea and a few days' work on a home computer can come up with one possible natural mechanism for global warming, how many more are out there waiting to be discovered? One might expect that the U.S. government would have put serious funding toward research into possible natural explanations for global warming before setting about to make hugely expensive and massively disruptive policy changes. But very little money has ever been awarded by the government for that purpose. Most funding for global warming research has gone toward building upon the assumption of anthropogenic climate change with increasingly complex computer models. And if you pay scientists enough money to find evidence of something, they will be happy to discover it for you. It is like the plot of an old movie where a corrupt police chief instructs all the detectives to pin a murder rap on the only suspect they have, based entirely on circumstantial evidence, and then tells the detectives that they will continue to get paid only if they succeed in doing so.

This is not the way unbiased scientific research should be carried out. But then, the study of global warming and climate change long ago lost any semblance of objectivity.

The bias exhibited by the Intergovernmental Panel on Climate Change should have been expected. The panel is first and foremost a political advocacy group formed to make the scientific case for manmade global climate change. One of the IPCC's original directors told me as much in the early 1990s, not long after the panel was established. There was never any serious interest in looking for natural causes of global warming. And the public is expected to believe the IPCC's proclamations and predictions because the physics of climate change are said to be too complicated for the average person to understand. You are to trust the experts to interpret the output of those complex models that run on the world's fastest computers.

It's not that we shouldn't be using computer models. I believe that mathematical models representing physical relationships are essential to an understanding of the complex processes that occur in nature. You don't really comprehend a physical process until you can demonstrate how things work with actual numbers put into equations representing those processes in action. And we have seen that a simple climate model – a single equation – can indeed help us understand what we observe in nature.

But the climate modelers with their supercomputers have missed the forest for the trees. As the famous modeler and cloud expert Robert Cess said many years ago, the models "may be agreeing now simply because they're all tending to do the same thing wrong. It's not clear to me that we have clouds right by any stretch of the imagination."[1] You seldom hear such an honest admission from the climate research community anymore. Climate scientists have taken for granted that global warming is manmade because no one could think of what else might be causing it.

The hubris of the claim that mankind now controls the climate system is astounding. I would think that the first place one should look to find explanations for climate change is in nature, not in the tailpipe of an SUV. The idea that the Earth can cause its own climate change seems entirely plausible to me as a meteorologist, and I have found that most meteorologists are distrustful of manmade explanations for global warming. Even the public understands that there is natural climate variability. It is the presumed experts – the climate modelers – who have rejected the concept of natural climate change.

One reason why I have written this book for the general public, rather than just let my scientific publications make my case, is that the climate research establishment has come to be paralyzed by groupthink. It has become inbred and has lost its objectivity.[2] Climate modelers ignore published science that does not fit their paradigm, in which humans rule the climate system. The tendency for scientists' objectivity to be compromised by their own (or their managers') policy preferences has been called "an insidious kind of scientific corruption."[3]

Moreover, it has now become next to impossible to publish research results that conflict with the IPCC's official line, partly because of the political muscle exercised by the IPCC and its supporters in government. Even our major science journals have unwritten editorial policies that prevent the publication of scientific results that cast doubt on the paradigm of anthropogenic climate change. No journal would admit to editorial bias, but many of us have learned that an editor can send a problematic manuscript to one alarmist reviewer who will poke enough holes in the science to allow the editor to reject it. I have even had peer reviewers tell me that my conclusions needed to be changed so as to conform with the IPCC's position. This is certainly putting the cart before the horse.

At the same time, any similar holes or weaknesses in papers that support the case for manmade global warming are conveniently ignored. The scientific hypocrisy never ceases to amaze me. If a study argues that there is likely to be even more global warming in the future than we have anticipated, that is considered an acceptable conclusion. And you can bet the news media will love it.

This leads to another reason why I wrote this book: The usual news outlets have taken on the role of censors, refusing to report any new science that does not accord with their worldview. When it comes to global warming, they have made sure that only certain kinds of scientific results are reported to you, the citizen. Thus the public has been kept in the dark about some of the research that their tax dollars have paid for – research which could have a huge impact on our understanding of the global warming issue, and on policy decisions related to energy use, cost, and availability.

So I decided to bypass the news media. I have tried to bridge the gulf between what the public understands on a conceptual level and the immensely complex models that the IPCC uses. I have peeled back the superfluous details of global warming theory to reveal the most basic components of global warming predictions. In contrast to the IPCC's magical mystery mega-models running on supercomputers, I have used a simple climate model to demonstrate these concepts. If you understand sunlight warming the inside of a car, you can understand *forcing*. If you understand that rolling the car window down will reduce that warming, you can understand *feedback*. These two concepts are all you really need in order to understand a simple yet powerful model of global average temperature change.

And it's something you really can try at home, kids. You can get your copy of the model at http://www.drroyspencer.com.

The evidence I have presented in this book strikes at the heart of the theory of manmade global warming. Feedbacks and climate sensitivity constitute the holy grail of climate research. If the sensitivity of the climate system is known, then we can easily calculate just how much warming will result from adding more greenhouse gases to the atmosphere. This is not just another piece in the global warming puzzle; it is what mostly determines how the finished puzzle will look.

As of 2009, it appears that the PDO may have switched back to its cooling phase, just as it did in the 1940s. Only time will tell whether Arctic sea ice continues its return to more "normal" levels and global temperatures continue refusing to climb. Maybe in a few years "global warming" will refer to a period of history in which humanity just overreacted, rather than to predictions of future global destruction.

Now we are at a bizarre point where carbon dioxide is considered a pollutant rather than a scarce nutrient that is necessary for life on Earth to survive. In fact, it might well be that more CO_2 in the atmosphere will be a good thing. During most of geologic history, there was much more CO_2 in the atmosphere than there is today. Life on land and in the ocean flourished, just as it does in greenhouses where the CO_2 content of the air is pumped up to three times the atmospheric concentration. The fear that we have instilled in people over having more CO_2 in the atmosphere is unwarranted. I predict that the latest and most uncertain concern over our CO_2 emissions, ocean acidification, will eventually turn out to be a false alarm, too.

The public has been misled by politicians and news reporters who have selectively filtered the science and economics related to climate change and energy use. I wanted to provide the interested public with a resource for cutting through the hype. I am betting that there are many scientists, physicists, chemists, engineers, and even economists out there who will be perfectly comfortable with what I have presented. They deal with similar concepts in their own lines of work. For an issue as important as global warming, with its major policy implications, there needs to be more grass-roots participation in the debate. It is simply too dangerous to allow the climate modelers to keep hiding behind their magic veil of complexity.

I want to stress again that I do not care where our energy comes from, and I am not paid by Big Oil to support their position. But I do care that the energy we use be as affordable and accessible as possible, to as many people as possible. Petroleum and coal executives will do whatever is in the best interests of their companies; they may even find it necessary to play along with any carbon tax or cap-and-trade scheme that the government imposes. But even if they do, I will continue to support the inexpensive energy we have now until alternative forms become cost-competitive. And since nature seems to enjoy more carbon dioxide in the atmosphere, I will continue to challenge those who demonize it.

I am under no illusion that this book will settle the scientific debate over the roles of mankind versus nature in global warming and climate change. Quite the opposite: I am hoping that the scientific debate will finally *begin*.

Notes

Introduction

1 Intergovernmental Panel on Climate Change, *Climate Change 2007: The Physical Science Basis* (New York: Cambridge University Press, 2007), <http://www.ipcc.ch/ipccreports/assessments-reports.htm>

2 Roy W. Spencer and William D. Braswell, "Potential Biases in Cloud Feedback Diagnosis: A Simple Model Demonstration," *Journal of Climate* 21 (2008): 5624–5628.

3 Roy W. Spencer, *Climate Confusion* (New York: Encounter Books, 2008).

4 Fred Lucas, "Al Gore's Carbon Empire: Cashing In on Climate Change," *Foundation Watch*, August 2008, <http://www.capitalresearch.org/pubs/pdf/v1217525953.pdf>

5 James Hansen, "Climate Change: On the Edge," *The Independent*, February 17, 2006.

6 Energy Information Administration, "What Will It Take to Stabilize Carbon Dioxide Concentrations?" *International Energy Outlook 2008*, June 2008, <http://www.eia.doe.gov/oiaf/ieo/scdc.html>

7 *Massachusetts et al. v. Environmental Protection Agency et al.* 05-1120 (April 2, 2007), <http://www.supremecourtus.gov/opinions/06pdf/05-1120.pdf>

8 Hugo Robinson and Neil O'Brien, "Europe's Dirty Secret: Why the EU Emissions Trading Scheme Isn't Working," *Open Europe*, August 2007, <http://www.openeurope.org.uk/research/etsp2.pdf>

9 Gregory Feifer, "Russia Pushes to Grow Gazprom's Reach, Control," NPR, January 6, 2009, <http://www.npr.org/templates/story/story.php?storyId=98874958>

10 Philip Mercer, "So, Al Gore, What's the One Thing We Can All Do to Tackle Climate Change?" *The Independent*, July 7, 2007, <http://www.independent.co.uk/environment/climate-change/so-al-gore-whats-the-one-thing-we-can-all-do-to-tackle-climate-change-456269.html>

11 Michelle Nichols, "Gore Urges Civil Disobedience to Stop Coal Plants," Reuters, September 24, 2008, <http://www.reuters.com/article/environmentNews/idUSTRE48N7AA20080924?sp=true>. And, *A Call to Action on Global Warming from Dr. James Hansen*, GreenPeace USA, video file, February 18, 2009, <http://www.youtube.com/watch?v=PPCFx1fMBeI&eurl=http://itsgettinghotinhere.org/2009/02/18/dr-james-hansen-calls-for-civil-disobedience-at-the-capitol-march-2nd/&feature=player_embedded>

12 *What Do Americans Believe about Climate Change?* The Nature Conservancy, October 2008, <http://www.nature.org/initiatives/climatechange/features/art26253.html>

13 "The FACTS about Our Changing Climate," panel discussion, AMS 36th Conference on Broadcast Meteorology, Denver, June 28, 2008.

14 Reto Knutti and Gabriele C. Hegerl, "The Equilibrium Sensitivity of the Earth's Temperature to Radiation Changes," *Nature Geoscience* 1 (2008): 735–743.

15 Roy W. Spencer, William D. Braswell, John R. Christy, and Justin Hnilo, "Cloud and Radiation Budget Changes Associated with Tropical Intraseasonal Oscillations," *Geophysical Research Letters* 34 (2007), doi:10.1029/2007GL029698.

16 Ellen Goodman, "No Change in Political Climate," *Boston Globe*, February 9, 2007, <http://www.boston.com/news/globe/editorial_opinion/oped/articles/2007/02/09/no_change_in_political_climate/>

Chapter 1 · Climate Change Happens

1 Craig Loehle, "A 2,000-Year Global Temperature Reconstruction Based on Non-Tree Ring Proxies," *Energy and Environment* 18 (2007): 1049–1058.

2 L. K. Barlow et al., "Interdisciplinary Investigations of the End of the Norse Western Settlement in Greenland," *The Holocene* 7 (1997): 489–499.

3 Brian M. Fagan, *The Little Ice Age: How Climate Made History, 1300–1850* (New York: Basic Books, 2001).

4 HadCRUT3 combined land-marine surface temperature dataset, available from <http://www.cru.uea.ac.uk/cru/data/temperature/>

5 J. R. Christy et al., "Error Estimates of Version 5.0 of MSU-AMSU Bulk Atmospheric Temperatures," *Journal of Atmospheric and Oceanic Technology* 20 (2003): 613–629.

6 Wang Hui-Jun et al., "El Niño and the Related Phenomenon Southern Oscillation (ENSO): The Largest Signal in Interannual Climate Variation," *Proceedings of the National Academy of Sciences* 96 (1999): 11071–11072, <http://www.pnas.org/content/96/20/11071.full.pdf+html>

7 Chris Newhall, James W. Hendley II, and Peter H. Stauffer, *The Cataclysmic 1991 Eruption of Mount Pinatubo, Philippines*, U.S. Geological Survey Fact Sheet 11397 (1997), <http://pubs.usgs.gov/fs/1997/fs113-97/>

8 Michael E. Mann, Raymond S. Bradley, and Malcom K. Hughes, "Global-Scale Temperature Patterns and Climate Forcing over the Past Six Centuries," *Nature* 392 (1998): 779–787.

9 Michael E. Mann and P. D. Jones, "Global Surface Temperature over the

Past Two Millennia," *Geophysical Research Letters* 30 (2003), doi:10.1029/2003GL017814.

10 Michael E. Mann et al., "Proxy-Based Reconstructions of Hemispheric and Global Surface Temperature Variations over the Past Two Millennia," *Proceedings of the National Academy of Sciences* 105 (2008): 13252–13257.

11 Intergovernmental Panel on Climate Change, "Summary for Policymakers," *Climate Change 2001: The Scientific Basis* (2001), <http://www.ipcc.ch/ipccreports/tar/wg1/005.htm>

12 Christopher C. Horner, *Red Hot Lies: How Global Warming Alarmists Use Threats, Fraud, and Deception to Keep You Misinformed* (New York: Regnery, 2008).

13 Gerald R. North et al., *Surface Temperature Reconstructions for the Last 2,000 Years* (Washington, D.C.: National Academies Press, 2006).

14 Stephen McIntyre and Ross McKitrick, "Hockey Sticks, Principal Components, and Spurious Significance," *Geophysical Research Letters* 32 (2005), doi:10.1029/2004GL021750.

15 Craig Loehle, "A Mathematical Analysis of the Divergence Problem in Dendroclimatology," *Climatic Change* (2008), doi:10.1007/s10584-008-9488-8.

16 Gilbert P. Compo and Prashant D. Sardeshmukh, "Oceanic Influences on Recent Continental Warming," *Climate Dynamics* 32 (2009): 333–342.

17 D. H. Douglass, J. R. Christy, B. D. Pearson, and S. F. Singer, "A Comparison of Tropical Temperature Trends with Model Predictions," *International Journal of Climatology* 27 (2007), doi:10.1002/joc.1651.

18 Deborah Balk et al., Global Urban-Rural Mapping Project (GRUMP), Socioeconomic Data and Applications Center, <http://sedac.ciesin.columbia.edu/gpw/>

19 Ross R. McKitrick and Patrick J. Michaels, "Quantifying the Influence of Anthropogenic Surface Processes and Inhomogeneities on Gridded Global Climate Data," *Journal of Geophysical Research – Atmospheres* 112 (2007), doi:10.1029/2007JD008465.

20 Edward N. Lorenz, "Deterministic Non-Periodic Flow," *Journal of the Atmospheric Sciences* 20 (1963): 130–141.

21 N. J. Mantua, S. R. Hare, Y. Zhang, J. M. Wallace, and R. C. Francis, "A Pacific Interdecadal Climate Oscillation with Impacts on Salmon Production," *Bulletin of the American Meteorological Society* 78 (1997): 1069–1079, <http://www.atmos.washington.edu/~mantua/abst.PDO.html>

22 PDO data are updated monthly at <http://jisao.washington.edu/pdo/PDO.latest>

23 V. Ramanathan et al., "Cloud-Radiative Forcing and Climate: Results from the Earth Radiation Budget Experiment," *Science* 243 (1989): 57–63, doi: 10.1126/science.243.4887.57.

24 C. E. P. Brooks, "The Warming Arctic," *Meteorological Magazine* 73 (1938): 29–31.

25 "Arctic Ocean Getting Warm; Seals Vanish and Icebergs Melt," *Washington Post*, November 2, 1922, p. 2: "great masses of ice have now been replaced by moraines of earth and stones . . . at many points well-known glaciers have entirely disappeared."

26 Historic Naval Ships Association, "RCMPV St. Roch," accessed March 2, 2009, <http://hnsa.org/ships/stroch.htm>

27 Southern Oscillation Index data since 1866 are updated monthly and are available at <http://www.cgd.ucar.edu/cas/catalog/climind/SOI.signal.ascii>

28 N. S. Keenlyside et al., "Advancing Decadal-Scale Climate Prediction in the North Atlantic Sector," *Nature* 453 (2008): 84–88.

29 Michael Reilly, "Global Warming: On Hold?" *Discovery News*, March 2, 2009, <http://dsc.discovery.com/news/2009/03/02/global-warming-pause.html>

Chapter 2 · "We Are Going to Destroy the Creation"

1 P. T. Doran and M. Kendall Zimmerman, "Examining the Scientific Consensus on Climate Change," *Eos, Transactions, American Geophysical Union* 90:3 (2009), doi:10.1029/2009EO030002.

2 "44% Say Global Warming Due to Planetary Trends, Not People," *Rasmussen Reports*, January 19, 2009, <http://www.rasmussenreports.com/public_content/politics/issues2/articles/44_say_global_warming_due_to_planetary_trends_not_people>

3 "Economy, Jobs Trump All Other Policy Priorities in 2009," *Pew Research Center Survey Report*, January 22, 2009, <http://people-press.org/report/485/economy-top-policy-priority>

4 Scott Rothschild, "NASA Climate Expert Warns Kansans of Dire Consequences of Global Warming," *Lawrence Journal-World*, September 23, 2008, <http://www2.ljworld.com/news/2008/sep/23/nasa_climate_expert_warns_dire_consequences_global/>

5 "Direct Testimony of James E. Hansen," before the State of Iowa Utilities Board, October 22, 2007, <http://www.columbia.edu/~jeh1/2007/IowaCoal_20071105.pdf>

6 Robin McKie, "President Has Four Years to Save Earth," *Guardian*, January 18, 2009, <http://www.guardian.co.uk/environment/2009/jan/18/jim-hansen-obama>

7 Paul Driessen, *Eco-Imperialism: Green Power, Black Death* (Bellevue, Wash.: Free Enterprise Press, 2003), 35–41.

8 Roy Innis, Alan Gottlieb, and Sean Hannity, *Energy Keepers, Energy Killers: The New Civil Rights Battle* (Bellevue, Wash.: Merril Press, 2008).

9 N. V. Vakulenko et al., "Evidence for the Leading Role of Temperature Variations Relative to Greenhouse Gas Concentration Variations in the Vostok Ice Core Record," *Doklady Earth Sciences* 397 (2004): 663–667.

10 Carl Wunsch, "Quantitative Estimate of the Milankovitch-Forced Contribution to Observed Quaternary Climate Change," *Quaternary Science Reviews* 23 (2004): 1001–1012.

Chapter 3 · Forcing: How Warming Gets Started

1 The rate at which infrared energy is given off by a "blackbody" emitter in watts per square meter is equal to the fourth power of the temperature (in Kelvin degrees) times the Stephan-Boltzman constant, $\sigma = 0.0000000567$ W m^{-2} K^{-4}. Earth is estimated to radiate energy like a blackbody at an average temperature of about 255 K, which when input into this relationship yields an average rate of infrared emission of about 240 W m^{-2}.

2 My estimate for the 3.5 percent contribution of CO_2 to Earth's greenhouse effect is based on the following calculation: The average surface temperature of Earth is estimated to be 288 K, but the equivalent temperature at which Earth emits infrared radiation to space is estimated to be about 255 K. This means that the greenhouse effect warms the surface of Earth by the difference, 33 deg. C. Next, radiative transfer model calculations generally yield 1.1 or 1.2 deg. C of warming for a doubling of atmospheric carbon dioxide. Therefore, the influence of a normal concentration of CO_2 in the atmosphere is about 1.15 divided by 33, which yields about 3.5 percent.

3 J. A. Hansen, A. Lacis, R. Ruedy, and M. Sato, "Potential Climate Impact of Mount Pinatubo Eruption," *Geophysical Research Letters* 19 (1992): 215–218, doi:10.1029/91GL02788.

4 CO_2 concentrations measured at Mauna Loa, Hawaii, are updated monthly and posted at <ftp://ftp.cmdl.noaa.gov/ccg/co2/trends/co2_mm_mlo.txt>

5 J. Hansen et al., "Earth's Energy Imbalance: Confirmation and Implications," *Science* 308 (2005): 1431–1435, doi:10.1126/science.1110252.

Chapter 4 · Feedback: How Much Warming Results from the Forcing

1 E. R. Thomas, G. J. Marshall, and J. R. McConnell, "A Doubling in Snow Accumulation in the Western Antarctic Peninsula since 1850," *Geophysical Research Letters* 35 (2008), doi:10.1029/2007GL032529.

2 R. S. W. van de Wal et al., "Large and Rapid Melt-Induced Velocity Changes in the Ablation Zone of the Greenland Ice Sheet," *Science* 321 (2008): 111–113.

Chapter 5 · How Mother Nature Fooled the World's Top Climate Scientists

1 Roy W. Spencer and William D. Braswell, "Potential Biases in Cloud Feedback Diagnosis: A Simple Model Demonstration," *Journal of Climate* 21 (2008): 5624–5628. And, Roy W. Spencer, William D. Braswell, John R. Christy, and Justin Hnilo, "Cloud and Radiation Budget Changes Associated with Tropical Intraseasonal Oscillations," *Geophysical Research Letters* 34 (2007), doi:10.1029/2007GL029698.

2 Reto Knutti and Gabriele C. Hegerl, "The Equilibrium Sensitivity of the Earth's Temperature to Radiation Changes," *Nature Geoscience* 1 (2008): 735–743.

3 G. L. Stephens, "Cloud Feedbacks in the Climate System: A Critical Review," *Journal of Climate* 18 (2005): 237–273.

4 Myles R. Allen and David. J. Frame, "Call Off the Quest," *Science* 318 (2007): 582–583.

5 Spencer and Braswell, "Potential Biases in Cloud Feedback Diagnosis."

6 Our global temperature data are updated monthly and are available at <http://vortex.nsstc.uah.edu/data/msu/t2lt/tltglhmam_5.2>

7 Bruce A. Wielicki et al., "Clouds and the Earth's Radiant Energy System (CERES): An Earth Observing System Experiment," *Bulletin of the American Meteorological Society* 77 (1996): 853–868.

8 The model equation for the time rate of change of temperature of the system is: $d\Delta T/dt = [F - \lambda \Delta T]/Cp$, where ΔT is the temperature departure from equilibrium in deg. C; F is radiative forcing in watts per sq. meter; λ is the total feedback parameter in watts per sq. meter per deg. C; and Cp is the total heat capacity of the system, which is dominated by the depth of the ocean involved in the temperature change. An Excel spreadsheet containing the model that you can run on your home computer is available at <http://drroyspencer.com>

9 P. M. Forster and K. E. Taylor, "Climate Forcings and Climate Sensitivities

Diagnosed from Coupled Climate Model Integrations," *Journal of Climate* 19 (2006): 6181–6194.

10 R. A. Madden and P. R. Julian, "Observations of the 40–50 Day Tropical Oscillation: A Review," *Monthly Weather Review* 122 (1994): 814–837.

11 Spencer, Braswell, Christy, and Hnilo, "Cloud and Radiation Budget Changes Associated with Tropical Intraseasonal Oscillations."

12 H. F. Hawkins and S. M. Imbembo, "The Structure of a Small, Intense Hurricane – Inez 1966," *Monthly Weather Review* 104 (1976): 418–442.

13 Spencer and Braswell, "Potential Biases in Cloud Feedback Diagnosis."

14 D. H. Douglass et al., "Thermocline Flux Exchange during the Pinatubo Event," *Geophysical Research Letters* 33 (2006), doi:10.1029/2006GL026355.

15 Forster and Taylor, "Climate Forcings and Climate Sensitivities Diagnosed from Coupled Climate Model Integrations."

16 Jia-Lin Lin et al., "Tropical Intraseasonal Variability in 14 IPCC AR4 Climate Models. Part I: Convective Signals," *Journal of Climate* 19 (2006): 2665–2690.

17 P. M. Forster and J. M. Gregory, "The Climate Sensitivity and Its Components Diagnosed from Earth Radiation Budget Data," *Journal of Climate* 19 (2006): 39–52.

Chapter 6 · Global Warming: Satellite Evidence for an Alternative Explanation

1 M. G. Dyck et al., *Polar Bears of Western Hudson Bay and Climate Change: Are Warming Spring Air Temperatures the "Ultimate" Survival Control Factor?* Science and Public Policy Institute, <http://scienceandpublicpolicy.org/images/stories/papers/reprint/polar_bear_sppi_word.pdf>

2 V. A. Semenov, "Influence of Oceanic Inflow to the Barents Sea on Climate Variability in the Arctic Region," *Doklady Earth Science* 418 (2008): 91–94, < http://resources.metapress.com/pdf-preview.axd?code=v22340616276763 1&size=largest>

3 Cynthia E. Sellinger et al., "Recent Water Level Declines in the Lake Michigan–Huron System," *Environmental Science & Technology* 42 (2008): 367–373, <http://pubs.acs.org/doi/full/10.1021/es070664%02B>

4 S. Levitus, J. Antonov, and T. Boyer, "Warming of the World Ocean: 1955–2003," *Geophysical Research Letters* 32 (2005): L02604 doi:10.1029/2004GL021592.

5 P. M. Forster and J. M. Gregory, "The Climate Sensitivity and Its Components Diagnosed from Earth Radiation Budget Data," *Journal of Climate* 19 (2006): 39–52.

6 P. M. Forster and K. E. Taylor, "Climate Forcings and Climate Sensitivities Diagnosed from Coupled Climate Model Integrations," *Journal of Climate* 19 (2006): 6181–6194.

7 D. H. Douglass and R. S. Knox, "Ocean Heat Content and Earth's Radiation Imbalance," *Physics Letters* A, 373 (2009): 3296–3300.

8 Isaac M. Held, "The Gap between Simulation and Understanding in Climate Modeling," *Bulletin of the American Meteorological Society* 86 (2005): 1609–1614.

Chapter 7 · CO_2: Dangerous Pollutant or Elixir of Life?

1 Robert T. Lackey, "Values, Policy, and Ecosystem Health," *BioScience* 51 (2001): 437–443.

2 A list of the published effects of elevated CO_2 on literally hundreds of species of plants is maintained at <http://co2science.org/data/plant_growth/plantgrowth.php>

3 Michael J. Behrenfeld et al., "Climate-Driven Trends in Contemporary Ocean Productivity," *Nature* 444 (2006): 752–755.

4 *Endangered Species of the Galapagos Islands*, Galapagos Conservancy Fact Sheet, <http://www.galapagos.org>

5 *The Missing Carbon Sink*, Woods Hole Research Institute, <http://www.whrc.org/carbon/missingc.htm>

6 M. Debora Iglesias-Rodriguez et al., "Phytoplankton Calcification in a High-CO_2 World," *Science* 320 (2008): 336–340.

7 Craig D. Idso, *CO_2, Global Warming and Coral Reefs: Prospects for the Future*, Center for the Study of Carbon Dioxide and Global Change, January 12, 2009, <http://co2science.org/education/reports/corals/coralreefs.pdf>

8 Terry Macalister, "Blow to Brown as BP Scraps British Renewables Plan to Focus on US," *Guardian*, November 7, 2008, <http://www.guardian.co.uk/business/2008/nov/07/bp-renewable-energy-oil-wind>

9 Terry Macalister, "Environment: Tar Sands–The New Toxic Investment," *Guardian*, September 17, 2008, <http://www.guardian.co.uk/environment/2008/sep/17/fossilfuels.carbonemissions>

10 Roy Innis, Alan Gottlieb, and Sean Hannity, *Energy Keepers, Energy Killers: The New Civil Rights Battle* (Bellevue, Wash.: Merril Press, 2008).

Chapter 9 · Summary & Conclusions

1 Richard A. Kerr, "Climate Change: Greenhouse Forecasting Still Cloudy," *Science* 276 (1997): 1040–1042, <http://www.sciencemag.org/cgi/content/full/sci;276/5315/1040>

2 Myanna Lahsen, "Seductive Simulations? Uncertainty Distribution around Climate Models," *Social Studies of Science* 35 (2005): 895–922.

3 Robert T. Lackey, "Normative Science," *Fisheries* 29 (2004): 38–39.

Index

A Note on the Type

The Great Global Warming Blunder *has been set in Nofret, a type designed by Gertrude Zapf von Hesse, the noted German calligrapher, type designer, and book artist. Strongly reminiscent of the designer's calligraphic hand, Nofret roman is beautiful at both text and display sizes. The italic is especially spirited and elegant, and together the types contribute a lively, contemporary energy to even the simplest page of text.*

DESIGN & COMPOSITION BY CARL W. SCARBROUGH